BIOHAZARD

BIOHAZARD

The Chilling True Story of the Largest Covert
Biological Weapons Program in the World—Told from
the Inside by the Man Who Ran It

KEN ALIBEK

WITH STEPHEN HANDELMAN

RANDOM HOUSE
NEW YORK

Library of Congress Cataloging-in-Publication Data
Alibek, Ken.
Biohazard: the chilling true story of the largest covert biological weapons program in the
world, told from the inside by the man who ran it/Ken Alibek with Stephen Handelman.
         p.   cm.
Includes bibliographical references and index.
ISBN 0-375-50231-9 (alk. paper)
1. Biological weapons—Soviet Union.   I. Handelman, Stephen.
II.Title.
UG447.8.A45   1998
358´.3882´0947—dc21   98-56454

Random House website address: www.atrandom.com

Printed in the United States of America on acid-free paper

6 8 9 7

Book design by Caroline Cunningham

[We are] determined for the sake of all mankind, to exclude
completely the possibility of bacteriological agents and
toxins being used as weapons;

[We are] convinced that such use would be repugnant to the
conscience of mankind and that no effort should
be spared to minimize this risk. . . .

—Preamble to the Biological and Toxin
Weapons Convention, 1972

CONTENTS

PROLOGUE

On a bleak island in the Aral Sea, one hundred monkeys are tethered to posts set in parallel rows stretching out toward the horizon. A muffled thud breaks the stillness. Far in the distance, a small metal sphere lifts into the sky then hurtles downward, rotating, until it shatters in a second explosion.

Some seventy-five feet above the ground, a cloud the color of dark mustard begins to unfurl, gently dissolving as it glides down toward the monkeys. They pull at their chains and begin to cry. Some bury their heads between their legs. A few cover their mouths or noses, but it is too late: they have already begun to die.

At the other end of the island, a handful of men in biological protective suits observe the scene through binoculars, taking notes. In a few hours, they will retrieve the still-breathing monkeys and return them to cages where the animals will be under continuous examination for the next several days until, one by one, they die of anthrax or tularemia, Q fever, brucellosis, glanders, or plague.

These are the tests I supervised throughout the 1980s and early 1990s. They formed the foundation of the Soviet Union's spectacular breakthroughs in biological warfare.

Between 1988 and 1992, I was first deputy chief of Biopreparat, the Soviet state pharmaceutical agency whose primary function was to develop and produce weapons made from the most dangerous viruses, toxins, and bacteria known to man. Biopreparat was the hub of a clandestine empire of research, testing, and manufacturing facilities spread out over more than forty sites in Russia and Kazakhstan. Nearly every important government institution played a role in the Soviet biological weapons program: the Ministry of Defense, the Ministries of Agriculture and Health, the Soviet Academy of Sciences, the Communist Party Central Committee, and, of course, the KGB. The System, as Biopreparat was often called, was more successful than the Kremlin had ever dared to hope.

Over a twenty-year period that began, ironically, with Moscow's endorsement of the Biological Weapons Convention in 1972, the Soviet Union built the largest and most advanced biological warfare establishment in the world. We were among the 140 signatories of the convention, pledging "not to develop, produce, stockpile or otherwise acquire or retain" biological agents for offensive military purposes. At the same time, through our covert program, we stockpiled hundreds of tons of anthrax and dozens of tons of plague and smallpox near Moscow and other Russian cities for use against the United States and its Western allies.

What went on in Biopreparat's labs was one of the most closely guarded secrets of the Cold War.

Before I became an expert in biological warfare I was trained as a physician. The government I served perceived no contradiction between the oath every doctor takes to preserve life and our preparations for mass murder. For a long time, neither did I.

Less than a decade ago, I was a much-decorated army colonel, marked out for further promotion in one of the Soviet Union's most elite military programs. If I had stayed in Russia, I would have been a major general by now, and you would never have heard my name. But in 1992, after seventeen years inside Biopreparat, I resigned from my position and fled with my family to the United States. In numerous debriefing sessions, I provided U.S. officials with their first comprehensive picture of our activities. Most of what I told them has never been revealed in public.

With the collapse of the Soviet Union, the danger once posed by our weapons work has sharply diminished. Biopreparat claims that it no longer conducts offensive research, and Russia's stockpile of germs and viruses has been destroyed. But the threat of a biological attack has increased as the knowledge developed in our labs—of lethal formulations that took our scientists years to discover—has spread to rogue regimes and terrorist groups. Bioweapons are no longer contained within the bipolar world of the Cold War. They are cheap, easy to make, and easy to use. In the coming years, they will become very much a part of our lives.

Since leaving Moscow I have encountered an alarming level of ignorance about biological weapons. Some of the best scientists I've met in the West say it isn't possible to alter viruses genetically to make reliable weapons, or to store enough of a given pathogen for strategic purposes, or to deliver it in a way that assures maximum killing power. My knowledge and experience tell me that they are wrong. I have written this book to explain why.

There are some who maintain that discussing the subject will cause needless alarm. But existing defenses against these weapons are dangerously inadequate, and when biological terror strikes, as I'm convinced it will, public ignorance will only heighten the disaster. The first step we must take to protect ourselves is to understand what biological weapons are and how they work. The alternative is to remain as helpless as the monkeys in the Aral Sea.

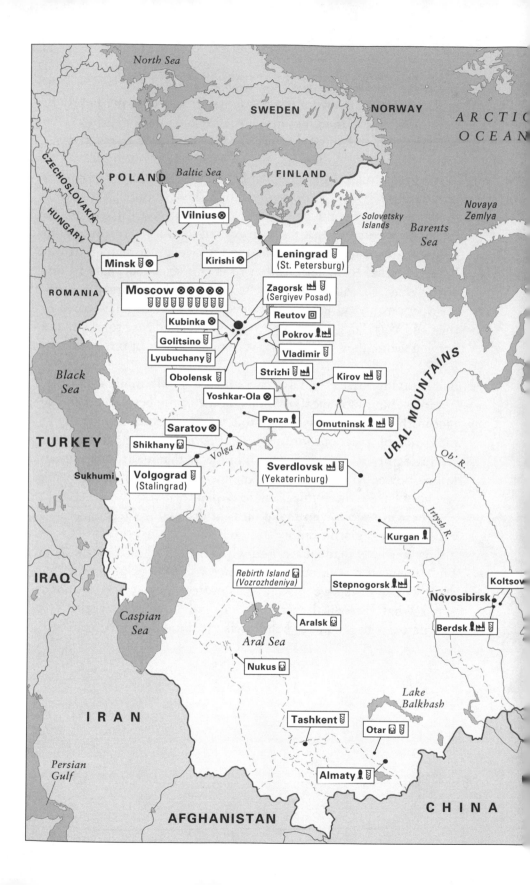

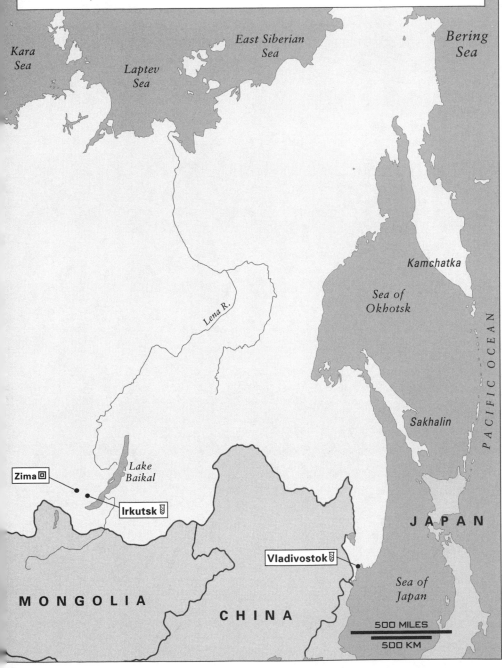

SOVIET BIOLOGICAL WARFARE INSTALLATIONS

- reserve mobilization facility
  (to be activated in time of war)
- biological weapon production
  facility
- biological weapon research
- testing grounds
- storage
- other

Kara Sea

Laptev Sea

East Siberian Sea

Bering Sea

Kamchatka

Sea of Okhotsk

PACIFIC OCEAN

Lena R.

Sakhalin

Zima

Lake Baikal

Irkutsk

JAPAN

Vladivostok

Sea of Japan

MONGOLIA

CHINA

500 MILES

500 KM

MILITARY

MEDICINE

1

ARMY HEADQUARTERS

Moscow, 1988

Late in the winter of 1988, I was called to a meeting at Soviet army headquarters on Frunze Street in Moscow. The note of urgency in the message was hard to ignore. "We've set aside a special room for you, Colonel," said the clipped voice on the phone.

A black Volga was waiting at the curb, its motor running. The two armed bodyguards who accompanied me on top-secret business were slouched alongside, their fur hats pulled low against the cold. One held the door open as I climbed into the backseat, and followed me inside. The second slid in beside my driver, Slava. I told Slava to drive quickly.

It was usually a thirty-minute drive across town from my office to Red Army headquarters, but a fresh snowfall that morning had turned the streets into an Arctic snarl of spinning tires and raging drivers. Once or twice the flashing blue light on our official vehicle aroused the attention of a traffic policeman, who thrust his gloved hand in the air to clear the way.

Close to an hour had passed by the time we finally pulled up in front of the austere granite building that housed the Ministry of

Defense. I entered through a side entrance and stamped the snow from my boots. A junior officer took me to a small adjoining room, where I was issued a pass, and then on to a guard booth, where a young soldier examined my pass and picture, stared hard, and waved me on.

The first officer led me up a flight of stairs to a heavy armored door with a coded lock. He punched in a series of numbers and we walked into the sprawling suite of offices occupied by the Fifteenth Directorate of the Soviet army, the military wing of our biological weapons program.

I unzipped my parka and tried to relax.

Although I was a colonel, I never wore my uniform. Like all military personnel at Biopreparat, I was provided with a cover identity as an ordinary scientist. I carried two different versions of my internal passport, the identity card required of every Soviet citizen. One identified me as a civilian employee of Biopreparat. The other showed my military rank.

I had moved to Moscow with my wife Lena and my three children a year earlier, in 1987, to take a position at Biopreparat headquarters. The move to the capital was a refreshing change from the dreariness of army life in the provinces.

Thirteen years at a succession of secret laboratories and institutes in some of the most remote corners of the Soviet Union had not prepared me for the bewildering pace of my new job. There were meetings every week at army headquarters, the Kremlin, the offices of the Communist Party Central Committee, or one of the myriad scientific institutes in our network. By the spring of 1988, when I was named first deputy chief, I was seeing a doctor for a stress-related illness.

The commander of the Fifteenth Directorate, Lieutenant General Vladimir Lebedinsky, looked at me disapprovingly when I entered his office. He was absorbed in a discussion with three colonels, none of whom I had seen before.

"It's about time," he said curtly.

I started to complain about the snow, the traffic, but he waved me into silence.

Of all the military commanders I dealt with, Lebedinsky was the one I most hated to keep waiting. He had taken a paternal interest in my career since we first met in a laboratory at Omutninsk, six hundred miles east of Moscow, where I'd been assigned for several years after graduating from military medical school. Then in his sixties and at the end of an illustrious military career, he was one of the few senior officers who didn't hold my youth against me. At thirty-eight, I had vaulted over older and more experienced scientists to become the youngest first deputy director in Biopreparat's history. Many of the scientists I used to work for were now taking orders from me, and they didn't bother to hide their resentment.

Lebedinsky turned to the three colonels.

"Are we ready?" he said.

They nodded, and the general led us into an adjoining soundproof room. Notepads had been placed on the large wooden table, in front of each chair.

An orderly arrived with four steaming glasses of tea. Lebedinsky waited for him to leave and firmly closed the door.

"I'm not staying," he said, as I glanced at the glasses and did a quick count.

The three colonels came from the Biological Group, a unit of the General Staff Operations Directorate whose role was to arm bombers and missiles with the weapons we produced. It was the first time I had met anyone from that unit. Biopreparat was then developing a new biological weapon every year. Most of our time was devoted to research; we paid little attention to the details of deployment.

Lebedinsky quickly explained the reason for the special meeting. A decision had been made at the highest levels, he said, to arm SS-18 missiles with disease agents.

"We need to calculate how much time it will take to prepare the missiles for launching. I'm counting on you to help us out."

I nodded, as if this were a perfectly reasonable request. But I had been caught off guard. The giant SS-18 missiles, which could carry ten five-hundred-kiloton warheads apiece over a range of six thousand miles, had never been considered before as delivery vehicles for a biological attack.

When the Soviet biological warfare program began in the 1920s, our scientists attached crop sprayers to low-flying planes and hoped that a contrary wind wouldn't blow the germs the wrong way. After World War II, bombers armed with explosives were added to the arsenal. The Cold War fueled the development of ever more destructive armaments, and by the 1970s we had managed to harness single-warhead intercontinental ballistic missiles for use in the delivery of biological agents. Multiple-warhead missiles represented more of a challenge. Few of the agents we had weaponized could be prepared in sufficient quantities to fill hundreds of warheads simultaneously.

Work I had done with anthrax a few years earlier must have caught the attention of our strategic planners. Through a series of tests, I'd found a way to create a more potent anthrax weapon, so that fewer spores would be needed in an attack. The new technique allowed us to load more missiles with anthrax without straining our labs' resources. In the language of American nuclear strategists, we could produce "more bang for the buck."

I was being asked to put my discovery to work.

The colonels knew little about the fine points of microorganisms, but they understood missile technology. If I could develop the pathogens in sufficient quantities, they would target the warheads on major cities in the United States and Europe.

I made a few quick calculations on my notepad. At least four hundred kilograms of anthrax, prepared in dry form for use as an aerosol, would be required for ten warheads.

Our seed stock for anthrax production was kept inside refrigerated storerooms at three production facilities in Penza, Kurgan, and Stepnogorsk. The seed stock would have to be put through a delicate fermenting process to breed the billions of spores required. The process was complicated—and it took time. A single twenty-ton fermenter working at full capacity could produce enough spores to fill one missile in one or two days. With additives, we could probably boost the output to five hundred or six hundred kilograms a day. I finished my calculations and leaned back in my chair.

"With the fermenters we have available, it would take ten to fourteen days," I said.

The colonels looked pleased. Two weeks was not a problem. No one expected to go to war overnight.

The colonels didn't tell me which cities had been targeted for biological attack, and I didn't ask. New York, Los Angeles, Seattle, and Chicago were some of the targets to come up in subsequent meetings, but they were abstract concepts to me at the time. All I cared about was ensuring that our weapons would do the job they were designed for.

We stood up to stretch. The tension in the room lifted. Three of us went out to the hallway for a smoke. I had discovered that you could learn more in such casual moments than in a month's worth of memos passed around The System. The colonels were suddenly talkative.

Pressure from the top military command was making their lives impossible, they complained. No sooner had one weapons system been organized than an order came down to refine another one.

I told them we were having the same problem—but we all read the newspapers. Mikhail Gorbachev and his team of self-described reformers were publicly heralding a new era of rapprochement with the West. We joked that the mysteries of *perestroika* were beyond the scope of simple military men.

I don't remember giving a moment's thought to the fact that we had just sketched out a plan to kill millions of people.

Anthrax takes one to five days to incubate in the body. Victims often won't know that an anthrax attack has taken place until after they begin to feel the first symptoms. Even then, the nature of the illness will not at first be clear. The earliest signs of trouble—a slight nasal stuffiness, twinges of pain in the joints, fatigue, and a dry, persistent cough—resemble the onset of a cold or flu. To most people, the symptoms will seem too inconsequential to warrant a visit to the doctor.

In this first stage, pulmonary anthrax can be treated with antibiotics. But it would take a highly alert public health system to recognize the evidence of an anthrax attack. Few physicians are trained to identify the disease, and the unremarkable nature of early symptoms makes an accurate diagnosis difficult.

The first symptoms are followed several days later by the anthrax "eclipse," a period in which the initial discomfort seems to fade, concealing the approaching danger. Proliferating bacteria will have begun to engulf the lymph nodes, local headquarters of the body's disease protection system. Within a matter of hours the bacteria will have taken over the entire lymphatic system. From there, they enter the bloodstream, continuing to multiply at a furious pace. Soon they begin to release a toxin that attacks all organs but is particularly damaging to the lungs, filling them with liquid and gradually cutting off their supply of oxygen.

Within twenty-four hours of this toxin's release, a victim's skin will begin to turn a faint bluish color. At this stage, every breath becomes more painful than the last. A choking fit and convulsions follow. The end usually comes suddenly: some victims of pulmonary anthrax have been known to die in the middle of a conversation. The disease is fatal in over 90 percent of untreated cases.

A hundred kilograms of anthrax spores would, in optimal atmospheric conditions, kill up to three million people in any of the densely populated metropolitan areas of the United States. A single SS-18 could wipe out the population of a city as large as New York.

Anthrax was not the only biological weapon earmarked for the SS-18s. When we sat down again after our break, we went over the available menu of toxic choices.

Plague could be prepared on a similar schedule. The plague weapon we had created in our laboratories was more virulent than the bubonic plague, which killed one quarter of the population of Europe in the Middle Ages. Smallpox was stockpiled in underground bunkers at our military plants, and we were developing a weapon prototype based on a rare filovirus called Marburg, a cousin of Ebola.

Nearly three hundred projects were outlined in the last Five-Year Plan we had been given by the Military-Industrial Commission, known by its Russian initials as the VPK. The VPK coordinated all of the Soviet Union's industrial production for military purposes. This gave it effective control over two thirds of the nation's industrial enterprises. A separate biological weapons directorate monitored our progress until our "products" were ready to be delivered to the Ministry of Defense, which we referred to as the Customer.

Our meeting ended after an hour or so of additional calculations. We shook hands, packed our papers, and congratulated one another on a productive session. On my way out I looked into Lebedinsky's office, but he was already gone. I never saw the colonels again.

Driving back to my office, I opened my briefcase to jot down a few more notes. Anyone who peered through the window would have seen a frowning, slightly overweight bureaucrat preoccupied with the country's business.

A strange twist of fortune had brought me to the pinnacle of power in Russia, a country that was not my own. My great-grandfather had been a *khan,* a member of the nobility in what is now Kazakhstan, in Central Asia, but I grew up in a system that lavished few of its privileges on non-Russians. My wife and daughter and two young sons had risen with me to a lifestyle inconceivable to the majority of Soviet citizens. With the combined salary of a senior bureaucrat and a high-ranking military officer, I earned as much as a Soviet government minister. But in the Communist system, money was not the measure of worth. What counted was the special status that gave us access to perks and influence in our supposedly egalitarian society.

Turning into the hidden driveway that led to the offices of Biopreparat on Samokatnaya Street, I began to focus on the rest of the day. I would only have time for a quick lunch before facing the mountain of messages and paperwork on my desk. The Volga glided past a concrete wall into a small courtyard. I packed up my notes and said a quick good-bye to Slava.

Slava never gave any hint of suspecting what went on in the meetings he took me to, and I never confided in him. We had been warned to be careful of what we said to lower-ranking employees. But I imagined he drew conclusions of his own, given the odd bits of conversation he overheard.

"Will you need me later?" he asked.

"Probably not till I go home," I told him. "I might be late again tonight."

The Moscow headquarters of Biopreparat, or the Main Directorate of the Council of Soviet Ministers as it was officially (and uninformatively) called, protected its secrets behind a yellow brick mansion with a green roof that had served as the home of the nineteenth-century vodka merchant Pyotr Smirnoff. The building's past and present associations provided an ironic symmetry: Smirnoff's product has done more than any foreign invader to undermine the health of Russian citizens.

Samokatnaya Street is so small and narrow that a pedestrian could easily miss it while walking down the nearby Yauza Embankment, overlooking one of the waterways that joins the Moscow River as it flows toward the Kremlin. There were five other buildings on our street, all largely obscured in the spring and summer by the thick foliage of ancient trees mercifully ignored by Communist city planners.

Despite its image as an impersonal city of cold buildings and wide boulevards, Moscow is dotted with hidden havens such as these. Even in winter, Samokatnaya Street was free from the surrounding bustle of the neighborhood, with its shabby residential apartment blocks, factories, and onion-dome churches.

Three centuries ago the area around Samokatnaya Street was known as the German Quarter. It was the only place in old Muscovy where foreigners (then universally described as German, regardless of their nationality) were allowed to live and carry on their business—at a safe distance from ordinary Russians, whom they might otherwise have infected with alien ideas, but close enough for the czars to exploit their skills.

A car bearing American diplomatic plates once turned up the street and parked opposite the building. KGB guards watched from inside as several people got out, peered at the fence for a few moments, and then returned and drove off. We talked about it for days afterward.

Savva Yermoshin, the KGB commander in charge of the building, was one of my closest friends at the time. He declared confidently that there was nothing to worry about, but security was tighter than usual for weeks.

I walked up a marble staircase, one of the few remaining architectural features of the old mansion, to my offices on the second floor. Nearly 150 people worked at headquarters, including technicians and administrative personnel, but the building exuded an air of restrained silence.

My secretary, Marina, was a plump, efficient woman in her late twenties. A slight tilt of her head told me that Yury Kalinin, the director of Biopreparat and my immediate boss, was already at work.

Marina sat with Kalinin's secretary, Tatyana, in the reception area connecting our offices. The two women disliked each other intensely because of some ancient quarrel and rarely spoke. When I wanted to speak to Kalinin, I had to address Tatyana directly. This time I bypassed her and knocked on his door. A brusque voice told me to enter.

Major General Yury Tikhonovich Kalinin, chief of the Main Directorate and deputy minister in the Ministry of Medical and Microbiological Industry, was sitting behind an enormous antique desk. A pair of heavy curtains had been drawn over the window near his armchair, and his office was wrapped in gloomy darkness. A picture of Mikhail Gorbachev hung on one wall. There was a gray safe in the corner.

I coughed and waited for him to notice me.

"So?" he said at last, without looking up.

"The meeting on Kirov Street lasted a little longer than I had expected," I said. "I thought I would check in."

"Interesting?" The general never used two words when one would do.

When I first visited his office as a young captain, Leonid Brezhnev's picture was hanging on the wall. Over the years, the portraits had changed to Yury Andropov, and then, briefly, to Konstantin Chernenko, reflecting the quick succession of ailing leaders who occupied the Kremlin during the early 1980s. Kalinin had no political opinions so far as I could tell. One leader was as good as another. What he respected was power.

I began to tell him about the plan to use SS-18s, but he seemed to know everything already. I wondered if Lebedinsky had called him.

"I knew you could handle it," he said and raised his hand in a gesture of dismissal. "Back to work, right?"

As usual, I was left with the impression that there were areas of this strange secret universe that I would never have access to. Not until much later did I realize that this was only Kalinin's way of spinning the illusions he needed to strengthen and maintain his authority.

Kalinin had risen swiftly in the army's chemical warfare corps—some claimed thanks to well-placed marriages—but he was an engineer, not a scientist. He was also impetuous, a man who enjoyed making quick decisions that took people by surprise—not the least his decision to bring me to Moscow. Against my natural inclinations, I admired him. In our gray bureaucracy, he stood out as an aristocrat.

He was tall, slim, an elegant dresser. His imported suits must have cost him more than he could afford, even on a general's salary. He lived with his second wife, a shy woman said to be the daughter of a four-star general, in a neighborhood Muscovites nicknamed "Tsarskoye Selo" ("Czar's Village")—a kind of inside joke because of the high-level officials it housed.

Kalinin never smoked and rarely drank, which set him apart from his peers, and was in excellent shape for a Soviet man in his early fifties. His black hair was always impeccably combed. With his high cheekbones and eagle nose, he looked like a member of the old Russian nobility.

Women adored him, and rumors of his amorous inclinations spiced up office gossip. Late one night I knocked on his door and walked in just as the general and Tatyana were hastily rearranging their clothes. He never mentioned it, and neither did I.

The charm Kalinin reserved for women was rarely experienced by his male subordinates. As I came to feel less awed in his presence, I would sometimes bring to him the case of a scientist or technician who needed a leave of absence for personal reasons. He invariably refused to listen.

"So," he would bark. "Now you're a psychiatrist!"

And he would order me back to work.

After even the briefest session with Kalinin I would retreat to my office with a sense of relief. I worked in a large room with a high ceiling and a window that looked out over a park by the riverbank. An oak desk I'd inherited from my predecessor occupied nearly half the space. The desk held the real symbols of my authority: five telephones. In Soviet government offices, an executive's status could be measured by the number of his phones—an indication of multiple sources of authority. I even had a *kremlyovka,* the small white phone that connects everyone in the upper reaches of the Soviet government, from the general secretary of the Communist Party to ordinary ministers of state.

Personal mementos of family or friends were taboo in the offices of senior government officials, but I had hung portraits of a few Russian scientists: D. I. Mendeleyev, who invented the periodic table of elements; Nikolai Pirogov, a nineteenth-century pioneer in military surgery; Professor Ilya Mechnikov, a Russian microbiologist who discovered cellular immunity.

I was eager to identify with Russia's glorious scientific past. Some day, I promised myself, I would return to pure research, or medicine.

The only other items in my office to suggest my training were books on microbiology, biochemistry, and medicine.

Sitting in a corner was a Western computer. I never used it, but it was another sign of "special" status in a regime that prohibited its citizens from owning a copier. I would have preferred a television or radio, but the KGB had banned them from the offices of senior personnel. Our security chiefs claimed that Western electronic surveillance was so good that foreign agents could decipher our deepest secrets by analyzing the vibrations of our conversations on glass. It made little sense to me: why not then ban the computer as well?

The KGB was thorough, and it lived by its own impenetrable logic. Once a month, security officers shooed all the lab chiefs and division heads out of their offices to check for bugs. Some believed that they were really checking on equipment they had themselves installed to record our conversations.

We all knew that we were being watched, but no one questioned the security precautions. We were engaged in secret combat against enemies who, we were told, would stop at nothing. The Americans had hidden behind a similar veil of secrecy when they launched the Manhattan Project to develop the first atomic bomb. Biopreparat, we believed, was our Manhattan Project.

Marina came in with a stack of messages. "Someone from Yermoshin's office is here to see you," she said.

A young KGB officer stepped in after her and waited for her to leave.

"Yes?" I said. But I knew what would happen next.

Since we operated under the fiction that none of the secretaries knew what we did, they could not be allowed in our presence when our "secrets" were discussed.

The officer handed me a folder with a note from Yermoshin. "Stuff from the third floor," I read in his hurried scrawl.

The third floor was home to our "First Department," the unit responsible for maintaining our secret files and all communications with Biopreparat facilities around the country. The only people allowed in, besides security personnel, were Kalinin and myself. It was administered by the KGB.

Sometimes I went upstairs myself. For one thing, it was the only place in the building where you could copy documents. The First Department was the sole custodian of our copier machine. It also offered a good opportunity to gossip with Yermoshin. Our families had spent time together a few weekends earlier.

I riffled through the papers in front of me while the officer stayed in the room, as he was obliged to do.

There were requests for supplies from one of our lab chiefs in Siberia; a notice of an "urgent" meeting at the Kremlin later that afternoon; a minor accident at one of our labs in western Russia which had sparked a debate between physicians at the Ministry of Health, who wanted to isolate the infected workers, and a general at the lab, who didn't. The general, typically, argued that isolation was unnecessary and would only stir up the staff. And there were the latest reports of a field test in the Aral Sea.

2

REBIRTH ISLAND

Aral Sea, 1982

Ten centuries ago, according to Russian legend, there was a mysterious kingdom on the shores of the Black Sea called Tmu Tarakan. Its name was variously translated as the "Place of Darkness" or the "Kingdom of Cockroaches." Modern-day Muscovites use the phrase whenever they want to describe a destination that is as loathsome as it is remote.

Every April during the 1980s and early 1990s a team of Biopreparat scientists set off for a place we jokingly referred to as Tmu Tarakan. Located twenty-three hundred miles south of Moscow, its name was Rebirth Island. Our teams would spend four or five months there, living in army barracks and testing that year's supply of biological weapons.

Rebirth Island is a tear-shaped speck in the Aral Sea, which divides the Central Asian countries of Uzbekistan and Kazakhstan. Languishing fifty miles off the Kazakh shoreline in waters so polluted by the runoff of agricultural fertilizers that nothing could possibly live in them anymore, it was the antithesis of its name.

The only year-round inhabitants as far as anyone could see were lizards.

Open discussion of the island's seasonal activities was strictly forbidden. Scientists couldn't tell their families where they were going, or why.

A half-dozen tumbledown buildings served as the scientific headquarters, as well as the barracks, for a migratory population that sometimes numbered as many as 150 people, including scientists, technicians, and a unit of soldiers responsible for firing the weapons and tying down the animals. A secret landing strip had been built nearby, but airplane traffic was kept to a minimum. When the first teams arrived in April, a thin layer of green grass covered the sandy soil. By June, the vegetation had withered to brown shoots. Winds swirling off the desert steppes provided the only respite from the heat. There were no birds, and the dust settled everywhere, getting into clothes, hair, and eyes, sweeping through the animal cages and into the food and scientists' notebooks.

The Aral Sea was once the world's fourth-largest inland body of water, but it has been shrinking every year since the 1960s when, in a wrong-headed agricultural experiment, Soviet state planners ordered the diversion of the Aral's river sources into concrete irrigation canals. The canals were to transform this part of Central Asia into a cotton bowl. After the first bumper harvests, the desert soil was exhausted, and local residents have been reaping the consequences ever since. The rivers silted over and clouds of toxic salts billow across the region every year, leaving one of the world's highest cancer rates in their wake.

We made a unique contribution of our own to the region's multiplying environmental tragedies.

In 1972, two fishermen died when a shift in the direction of the wind sent a cloud of plague over their boat. In the 1970s and 1980s, abnormally high incidences of plague were detected among rodents in inhabited areas north of the testing ground. Following the Soviet collapse in late 1991, doctors reported outbreaks of plague in several areas of Central Asia. It is impossible to prove that these outbreaks were connected to our activities, but it seems more than likely.

The army's Fifteenth Directorate, which ran the Rebirth Island

complex, operated a year-round command post in Aralsk, the closest mainland community. A single, nearly impassable dirt road
linked Aralsk with the outside world. The town was once a fishing
port, but the shrinking sea left it stranded like a shipwreck sixty
miles from shore. Once home to several thriving fish canneries, Aralsk had begun to shrivel up, following the pattern of the sea from
which it took its name.

We used to say that the most fortunate inhabitants of the Soviet
Union were the condemned monkeys of Rebirth Island. They were
fed oranges, apples, bananas, and other fresh fruits rarely seen by
Soviet citizens. Our work teams could only admire this vision of
plenty from afar. Each piece of fruit was carefully inventoried and
guarded to dissuade members of the scientific teams from giving in
to temptation. It was grudgingly acknowledged that our test subjects needed to stay healthy until their last breath, while the scientists, who had to subsist on rations of cold porridge and fatty
sausage, were expendable.

Our team members were better off than those who spent their
lives in the area. Scientists on the occasional foraging trips into
town were shocked to see the earthen huts with no running water
that served as dwellings for most of the inhabitants. Malnutrition
and hepatitis were common.

This was a familiar scene to anyone who has ever traveled in
the rural areas of the former Soviet Union, but it never failed to in-
furiate me. I was born several hundred miles away, on the site of
another failed agricultural experiment in southern Kazakhstan. All
Kazakhs know that the money spent on Soviet military programs
could have fed and clothed hundreds of communities like Aralsk.
But in those days, we would never have dared say such a thing.

When the day's experiments were completed, the small migrant
population of scientists and soldiers could only look forward to
nights of interminable boredom. Once or twice a week, sentimen-
tal Soviet war movies were shown on a rickety movie projector
powered by the camp's sole electric generator. Drinking was the
most popular social activity on the island. Although no vodka was
available, some enterprising souls obtained bottles of distilled spirits. Some took their solace a bit too seriously: alcoholism was a
chronic problem among the scientists on these expeditions.

Sex was the second most popular activity. Scientists rarely mixed with the soldiers, mostly young recruits, but the Biopreparat teams often included female technicians. The combination of loneliness and boredom fueled numerous affairs, as well as gossip, which spiced up the dry reports we received back in Moscow. The end of a testing period would inevitably bring news of a divorce or a pregnancy that would be hard to explain back home.

The assignment suited those who enjoyed enforced vacations from their wives, mistresses, or children, but for most people the blend of relentless monotony and sleepless vigilance made even the stressful conditions in our labs seem attractive. In Moscow all that counted was the steady stream of reports that gave our bureaucracy its principal justification.

At the height of the U.S. offensive biological weapons program, American scientists restricted themselves to developing armaments that could be countered by antibiotics or vaccines, out of a concern for protecting troops and civilians from potential accidents. The Soviet government decided that the best agents were those for which there was no known cure. This shaped the entire course of our program and thrust us into a never-ending race against the medical profession. Every time a new treatment or vaccine came to light somewhere, we were back in our labs, trying to figure out how to overcome its effects.

Trafficking in germs and viruses was legal then, as it is today. In the name of scientific research, our agents purchased strains from university research laboratories and biotech firms around the world with no difficulty. Representatives of Soviet scientific and trade organizations based in Europe, as well as in Africa, Asia, and Latin America, had standing instructions to look out for new or unusual diseases. It was from the United States, for instance, that we obtained Machupo, the virus that causes Bolivian hemorrhagic fever. We picked up Marburg, related to the Ebola virus, from Germany.

The KGB was our most dependable supplier of raw material. They were known within Biopreparat by the code name "Capturing Agency One." Vials arrived in Russia almost every month with exotic fluids, powders, and cultures gathered by our intelligence agents in every corner of the globe. They were then sent by diplomatic pouch to Moscow, where Biopreparat technicians cautiously repackaged them. When I worked in provincial institutes, I was often ordered to pick up these toxic dispatches with a pair of armed guards in tow.

We were never permitted to travel by air. The consequences of a crash in one of our aging Aeroflot planes was too horrible to contemplate. Instead, dressed in civilian clothes, we returned in the cramped, sweaty passenger compartments of trains, trying to be inconspicuous.

By the mid-1980s, every Biopreparat laboratory, scientific institute, and production facility was working at full capacity. There were new agents, new strains of viruses and bacteria, and new methods of dispersal to test every month. We even explored AIDS and Legionnaire's disease. Both, as it happened, proved too unstable for use on the battlefield or against civilian populations. After studying one strain of the AIDS virus collected from the United States in 1985, we determined that HIV's long incubation period made it unsuitable for military use. You couldn't strike terror in an enemy's forces by infecting them with a disease whose symptoms took years to develop.

We had greater success in our work on more traditional killers.

One of the most infectious diseases known to man, smallpox, was declared eradicated by the World Health Organization in 1980. The last naturally occurring case was reported in 1977, and the medical profession judges the minor health risk associated with a vaccine greater than forgoing inoculation. Today, it isn't possible to get a smallpox vaccination in the United States, unless you are a lab scientist or a member of the military. This was for us an excellent reason to weaponize it. Although we officially had a small amount of the virus in the Ivanovsky Institute of Virology in Moscow—matching the world's only other legal repository of the strain in the United States—we cultivated tons of smallpox in our secret lab in Zagorsk (now Sergiyev Posad), the famous Russian cathedral city a half hour's drive from the capital. At Zagorsk, we experimented with the culture until we came up with a weapons-quality variant. Smallpox was then quietly added to our arsenal.

By the 1980s, so many different varieties of unconventional

weapons were being developed and tested in the Soviet Union that a complex code, arranged according to letters of the alphabet, was developed to keep track of them. Words beginning with *F*, for example, were assigned to chemical weapons ("Foliant") and to psychotropic, or behavior-altering, biological and chemical agents ("Flute").

The letter *L* covered bacterial weapons. In order to further conceal what we were working on, each disease agent carried its own subcode. Plague was L1; tularemia L2; brucellosis and anthrax were L3 and L4, respectively. Glanders was L5, melioidosis L6, and so on. Weapons based on viruses fell under the letter *N*. Smallpox, for instance, was described in clandestine communications as N1. Ebola was N2, Marburg N3, Machupo or Bolivian hemorrhagic fever N4.

The skittish behavior of microorganisms leads many experts to question their effectiveness as weapons. One of the problems has always been to find a reliable means of delivery, one that prevents biological agents from losing virulence when they are dispersed. It is the equivalent of what biologists call a "vector" for the transmittal of disease.

Over the centuries, armies have often used primitive methods to spread pestilence. The Romans dropped poison into wells to contaminate their enemies' water supplies. The English gave blankets smeared with smallpox to Indians in the eighteenth century during the French and Indian Wars. Confederate troops in the American Civil War left corpses of animals to rot in ponds along the path of Union forces. And during World War II, Japanese planes dropped porcelain bombs containing billions of plague-infected fleas over Manchuria.

The most effective way of contaminating humans is through the air we breathe, but this has always been difficult to achieve. Soviet scientists combined the knowledge gained from postwar biochemistry and genetic research with modern industrial techniques to develop what are called "aerosol" weapons—particles suspended in a mist, like the spray of an insecticide, or a fine dust, like talcum powder.

Temperature and weather conditions will determine the success of an aerosol's dissemination. Bacteria and viruses are generally vulnerable to sunlight; ultraviolet light kills them quickly. Heavy rain or snow, wind currents, and humidity impede their effectiveness.

Such obstacles complicate the planning of a biological attack, but they are not insurmountable. A bioweaponeer will know to strike at dusk, during periods when a blanket of cool air covers a warmer layer over the ground—a weather condition called an inversion, which keeps particles from being blown away by wind currents. We packed our biological agents in small melon-sized metal balls, called bomblets, set to explode several miles upwind from the target city. Meticulous calculation would be required to hit several cities at the same time with maximum effectiveness, but a single attack launched from a plane or from a single sprayer perched on a rooftop requires minimal skill.

Primitive aerosols lose their virulence and dissipate quickly. In our labs, we experimented with special additives to keep our agents from decaying when transported over long distances and to keep them alive in adverse weather conditions. These manipulated agents, more stable and more lethal, were our biological weapons.

We would first test our aerosols in special static chambers, where air flow was controlled to monitor the particle distribution after a small explosion or discharge. The last stage in determining a weapon's efficiency was live animal tests, such as the ones we conducted in the Aral Sea.

We tested a variety of animals, including rabbits and guinea pigs, but monkeys, whose respiratory systems are so similar to ours, were the most effective surrogates for humans. An average person takes in ten liters of air every minute. A monkey inhales four. If four particles of an agent in a given volume of air killed at least 50 percent of the monkeys exposed to an aerosol, we could assume that ten particles would have an equally lethal effect on human beings.

Our standard measure of success for a biological weapon was referred to as Q_{50}, representing the amount needed to infect 50 percent of all exposed human beings in one square kilometer of territory. The Soviet Union devoted an enormous amount of time and money to developing concentrated aerosols that could reach the Q_{50} level with minuscule numbers of bacterial cells or viral particles.

The most effective biological weapons go on killing long after they are used. Some viruses, such as Marburg, are so hazardous that casually inhaling as few as three microscopic viral particles several days after an attack would be enough to kill you. Biowarfare strategists often look beyond the immediate target to focus on the epidemic behavior of disease-causing agents.

Unlike nuclear weapons, which pulverize everything in their target area, biological weapons leave buildings, transportation systems, and other infrastructure intact. They should properly be called mass casualty weapons, not weapons of mass destruction.

Until General Yury Tikhonovich Kalinin took over Biopreparat in 1979, six years after it was created by a secret Kremlin decree, the agency didn't achieve much. Its first chief was an uninspiring but pleasant army general named Vsevolod Ogarkov, who was transferred from the Fifteenth Directorate, the branch of the defense ministry that had supervised the development of biological weapons since World War II.

The commanders of the Fifteenth Directorate thought of Biopreparat as an extension of the military research program and therefore subject to their control. Ostensibly operating as a civilian pharmaceutical enterprise, the agency could engage in genetic research without arousing suspicion. It could participate in international conferences, interact with the world scientific community, and obtain disease strains from foreign microbe banks—all activities which would have been impossible for a military laboratory.

Conflict between the Fifteenth Directorate and Biopreparat was inevitable. The military hierarchy was not equipped to deal with the relatively free-wheeling atmosphere of scientific research at the new agency. Many of the colonels and generals who had crossed town from army headquarters to Samokatnaya Street were scientists themselves, and the simple act of shedding their military uniforms and donning civilian clothing liberated them overnight. Excited by the prospect of cutting-edge research, some chose to reject the strict parameters laid down by defense headquarters.

The army chiefs responded by doing everything in their power to undermine the agency, which they had come to view as an upstart child. Ogarkov, already overwhelmed by the bureaucratic hassles of establishing an untested new structure within the Soviet government, didn't have the strength to fight them. In 1975, Biopreparat was assigned a Five-Year Plan to develop new biological weapons. Four years later, not a single new weapon had been produced.

Few expected Kalinin, then a forty-one-year-old engineer trained at the army's chemical warfare academy, to make a difference. He was considered a dark horse from the beginning: his knowledge of biological weapons was limited, and he had few friends in the Fifteenth Directorate. But the elegant Kalinin proved a master of political intrigue, maneuvering himself from an obscure post as lab chief in Zagorsk into a management job at one of Biopreparat's institutes and then into the director's chair. A first marriage to the daughter of a lab director, and a second to the daughter of a senior general, helped smooth his path up the ladder. His knack for forging friendships with senior military commanders and academicians at the Soviet Academy of Sciences was equally useful.

Unable to challenge his opponents at army headquarters directly, Kalinin expanded his new dominion by stealth. With the help of powerful friends, Biopreparat's new chief seized institutes controlled by other agencies, grabbing their scientists and recruiting thousands of new ones. He obtained funds to erect dozens of research and production buildings where none had existed before. Between 1975 and 1980, the number of Biopreparat employees quintupled. Most of this increase came in the year after Kalinin's appointment.

Kalinin knew his empire-building was futile unless he could show results. With no significant weapons project under way after nearly two years at Biopreparat's helm, the general was, understandably, in a desperate mood.

In June 1981, he called me at the laboratory in Omutninsk, where I had just been reassigned as chief of technological development. I snapped to attention when I heard his voice. We were all in awe of Kalinin by reputation, and the sheer improbability of a call from the Great Man to a young scientist made me wonder what I could have done wrong.

"I want you to come to Moscow," he announced.

"Yes, of course," I replied.

"I'm going to nominate you deputy director of Omutninsk."

I knew the general had a mercurial temperament, but this sounded alarmingly whimsical. I was just six years out of graduate school, a thirty-one-year-old captain with a lot of energy and only a few achievements to show for it. My name had come to the attention of superior officers thanks to a technique I had recently developed for improving biological weapons production. But I had only just been promoted to my new job. I was nervous, and wanted at the time to say no.

"Well?" Kalinin was still waiting.

"I'll be there tomorrow morning," I said.

It was my first visit to Kalinin's office. I climbed to the second floor and stopped by his secretary's desk. She offered me a glass of tea.

"They're not ready for you yet," she said a little hesitantly. "There's a bit of a disagreement."

I could hear it. The shouting penetrated beyond Kalinin's closed doors, though I couldn't make out who was there or what they were saying.

Seconds later, a red-faced man came barreling out of the door and stopped in front of me. He looked me up and down.

"I don't know what you think you're up to!" he barked. "You're nothing but a puppy."

He stormed back into the office.

I waited another half hour or so, and then Kalinin himself came out. He looked mildly apologetic.

"Go to your hotel," he said. "Have some lunch. I'll call you later."

I did as I was told, but I couldn't eat. Privately, I'd decided to return to Omutninsk as soon as I diplomatically could and to pretend none of this had ever happened.

Kalinin called me late in the afternoon.

"Congratulations," he said. "You are now the new deputy director of Omutninsk."

I began to stammer out a reply, but he interrupted me in midsentence. "Just get down here at once."

That afternoon, Kalinin was in one of his rare expansive

moods. He took some pleasure in telling me what had happened. It was partly a way of flaunting his achievement before an admiring subordinate, but I'm sure he also wanted to let me know from the start who was boss.

"The thing is," he began, "a couple of the armchair generals in that room disagreed with your appointment—Benetsky, most of all."

Benetsky was Kalinin's deputy, and the man who had given me the tongue-lashing. He was also a powerful military bureaucrat who had recently been transferred from the Ministry of Defense. Everyone knew that Kalinin feared him.

"Benetsky insisted that a thirty-year-old captain can't possibly manage lieutenant colonels and majors," Kalinin said. "He said he had never heard of anything so ridiculous in all of his career."

Kalinin now had a sly smile on his face, a smile I would get to know well. "But I managed to convince him you'd do all right."

"How?" I ventured fearfully.

"You'll turn our tularemia project around."

It was an assignment no scientist of my age and experience could have expected to get so early in his career. Both Biopreparat and the Fifteenth Directorate had been searching for years for a way of making tularemia into a more effective biological weapon. I knew the project was fraught with risk, but I was caught up in the challenge.

Tularemia is a debilitating illness, rife among wild animals and common in the Rocky Mountains, California, Oklahoma, parts of eastern Europe, and Siberia. It is a hardy organism, capable of surviving for weeks, sometimes months, in decaying animal corpses. Tularemia is primarily transmitted to humans by ticks, mosquitoes, and wild rabbits, though squirrels, sheep, cats, and dogs have also been identified as carriers. While highly infectious, it almost never spreads directly from one person to another.

Victims can be laid up for weeks with chills, nausea, headaches, and fever. If left untreated, symptoms usually last two to four weeks, but they can continue for months. *Francisella tularensis* is lethal in 30 percent of untreated cases.

After World War II, scientists in the United States, Britain, and

Canada developed tularemia for use on the battlefield, where it could immobilize an entire division through the intensive medical care required for each stricken soldier.

Soviet commanders considered tularemia an unpredictable weapon for close-quarter tactical maneuvers. The risk of infecting one's own troops was high. But we had obtained, from a leading international research institute in Europe, a strain capable of overcoming immunity in vaccinated monkeys.

Our civilian credentials ensured that no questions were asked when we requested it. As far as we knew, there had never been an attempt anywhere in the world to weaponize a vaccine-resistant strain of tularemia. For Kalinin, the project represented a chance to prove what Biopreparat could do.

We spent months in frustrating calculations and false starts, but by the early summer of 1982, we were ready to test our new weapon on Rebirth Island. The military had been running biological weapons trials there for years, but this was the first time Biopreparat would use the testing grounds on the Aral Sea—we'd never had a weapon to test before. We knew the high command would be watching jealously and counting on us to fail.

The process of getting a biological weapon approved for inclusion in the Soviet arsenal had changed little since the war. Test results had to be vetted by officials at the Military-Industrial Commission in Moscow and the complex lab work involved in preparing what we called the "final formulation"—the liquid or powdered version of the agent to be used in bombs or sprayers—had to be written down like a recipe, so that it could be reproduced by any technician at any of our production plants.

If the Ministry of Defense was satisfied with the test results, and if the final formulation was judged to be sound, a report would be prepared for the chief of the Army General Staff, who would issue an order officially designating the new weapon part of the Soviet arsenal. The recipe would then be stamped "top secret" and filed at headquarters, with a copy to the production plant assigned to produce the weapon.

If the Ministry of Defense was not satisfied, research would begin all over again.

Five hundred monkeys were ordered from Africa for tularemia tests on Rebirth Island. We scheduled a series of special flights from the military airport at Kubinka, outside Moscow.

Importing that many animals at one time without arousing suspicion was not as difficult as it sounds. The arrangements were made well in advance by clandestine overseas trading associations run by the Soviet Ministry of External Trade, which also supplied us with the cages and other special equipment. I don't know how we would have explained such an emergency request if we had been asked, but in the old Soviet Union, no one ever asked.

Since we were testing a vaccine-resistant weapon, all of the monkeys had to be immunized before they were exposed.

Back in the labs at Omutninsk, we filled twenty bomblets with my new tularemia weapon and prepared them for shipment to Rebirth Island.

Two senior officers were placed in charge of the tularemia test team that year: General Anatoly Vorobyov, first deputy director of Biopreparat, and General Lebedinsky of the Fifteenth Directorate. I was ordered to stay at Omutninsk to prepare an alternate formula for testing later that summer. I found it hard to concentrate on my lab work.

Information from Rebirth Island was difficult to come by. There was no telephone at the test center, and the only communication with Aralsk was through cryptograms on the army's closed-circuit communications network. I would have no idea what had really happened until my colleagues made their way back to Omutninsk.

When the coded test results filtered back the news was better than anyone had expected, including me. Nearly all the immunized monkeys died. I received a call from a very happy Kalinin.

"Kanatjan!" he shouted over the phone line from Moscow. It was one of the few times in those early days that he used my first name. "You are a great man!"

Other congratulatory calls followed from colleagues in Moscow who had heard about the results. A few weeks later I made the long trip to Moscow again, this time to receive a special military medal for "wartime services" from the gloating Kalinin.

Perplexingly, there was no word from the Fifteenth Directorate for weeks—not even an indication that a formal review was under way. Then we received a stilted letter from the Ministry of Defense.

"This weapon cannot be accepted into the arsenal," the letter said. "Our investigation has shown that preliminary testing of blood samples in affected monkeys was not done correctly."

The ministry was right. General Vorobyov had neglected a couple of established procedures in his haste to prepare the monkeys. His lapse of attention had little bearing on the success of the tests, but the military decided to put us in our place. The Fifteenth Directorate was not about to hand its rival an easy victory, particularly on a project directed by a "puppy." Kalinin was furious.

It turned out to be a minor setback. The next year we conducted new tests with an even more efficient dry variant of tularemia, following all the procedures meticulously, and the new version of weaponized tularemia entered the Soviet arsenal. The achievement launched Biopreparat as a significant force in the nation's weapons establishment. Kalinin was now safely ensconced in the backstairs club of Kremlin military politics, and I had been initiated into the fraternity.

Meanwhile, at Rebirth Island, everything connected to the tularemia test, from research notes to blood samples to the monkeys' corpses, had to be incinerated. The testing area was swept clean of all signs of human and animal occupation and then disinfected to eliminate all "footprints" of biological activity.

Open-air testing at Rebirth Island stopped in 1992. Records of what happened there no longer exist.

3

MILITARY MEDICINE

Stalingrad, 1942

Biological warfare was the furthest thing from my mind when I entered the military faculty of the Tomsk Medical Institute in 1973 to begin graduate studies as a cadet intern. I was planning to become a military psychiatrist, until a professor gave me an assignment that changed my relationship to medicine. He asked me to analyze a mysterious outbreak of tularemia on the German-Soviet front shortly before the Battle of Stalingrad in 1942.

The assignment was for a course in epidemiology. Most students disliked the professor, the balding, stony-faced Colonel Aksyonenko, but I respected him. He didn't seem to take himself quite so seriously as the other members of our faculty, who flaunted their advanced degrees and senior ranks. I even liked his lectures. They captured my attention in a way that other courses in military medicine, field surgery, and hygiene—courses we had to take before receiving an officer's commission and a posting—never did.

At the institute library, I spent several nights leafing through the twenty-five-volume *History of Soviet Military Medicine in the Great Patriotic War: 1941–1945* and pulled from the shelves dust-caked scientific journals from the wartime era. As I read, I became fascinated by what seemed an inexplicable sequence of events.

The first victims of tularemia were German panzer troops, who fell ill in such large numbers during the late summer of 1942 that the Nazi campaign in southern Russia ground to a temporary halt. Thousands of Russian soldiers and civilians living in the Volga region came down with tularemia within a week of the initial German outbreak. The Soviet high command rushed ten mobile military hospitals into the area, a sign of the extraordinary rise in the number of cases.

Most of the journals reported this as a naturally occurring epidemic, but there had never been such a widespread outbreak in Russia before. One epidemiological study provided a telling statistic: in 1941, ten thousand cases of tularemia had been reported in the Soviet Union. In the year of the Stalingrad outbreak, the number of cases soared to more than one hundred thousand. In 1943, the incidence of disease returned to ten thousand.

It seemed strange that so many men had first fallen sick on one side only. The opposing armies were so close together that a simultaneous outbreak was all but inevitable. Only exposure to a sudden and concentrated quantity of tularemia could explain the onslaught of infections in the German troops alone. Seventy percent of those infected came down with a pneumonic form of the disease, which could only have been caused by purposeful dissemination.

When I walked into my professor's office with a draft of my paper, I thought I had solved the puzzle. He was concentrating on the latest edition of *Krasnaya Zvezda,* the official army newspaper.

"So, what have you discovered?" Aksyonenko asked, smiling up at me before returning to his paper.

"I've studied the records, Colonel," I said cautiously. "The pattern of the disease doesn't suggest a natural outbreak."

He looked up sharply. "What does it suggest?"

"It suggests that this epidemic was caused intentionally."

He cut me off before I could continue.

"Please," he said softly. "I want you to do me a favor and forget you ever said what you just did. I will forget it too."

I stared at him in confusion.

"All I asked you to do was describe how we handled the outbreak, how we contained it." Aksyonenko had begun to frown. "You have gone beyond your assignment."

He pointed to the paper I had left on his desk.

"I don't want to see this until you've given it some thought. And never . . . never mention to anyone else what you just told me. Believe me, you'll be doing yourself a favor."

The final version of my paper did not mention the likelihood of deliberate infection. Yet Aksyonenko's reaction convinced me that I was on to something: Soviet troops must have sprayed tularemia at the Germans. A sudden change in the direction of the wind, or contaminated rodents passing through the lines, had infected our soldiers and the disease had then spread through the region.

Years later, an elderly lieutenant colonel who worked in the secret bacteriological weapons facility in the city of Kirov during the war told me that a tularemia weapon was developed in Kirov in 1941, the year before the Battle of Stalingrad. He left me with no doubt that the weapon had been used.

The lesson of Stalingrad would not be forgotten by our biological warfare strategists. In the postwar years, the Soviet high command shifted its attention from battlefield deployment to "deep targets" far behind enemy lines, where there was no danger of infecting one's own troops.

Stalingrad was a test of survival for the Soviet Union. If the city had been lost, the nation's industrial heartland in the Urals would have fallen before the advancing German tanks. More than one million of our soldiers died defending the city. In forcing the Germans into a humiliating retreat, they had turned the tide of the war.

The moral argument for using any available weapon against an enemy threatening us with certain annihilation seemed to me irrefutable. I came away from that assignment fascinated by the notion that disease could be used as an instrument of war. I began to read everything I could find about epidemiology and the biological sciences.

Near the army barracks on Rebirth Island stands a grave marked by a small headstone with an illegible name. A young woman was

buried there, a member of one of the first teams of army scientists
to conduct weapons trials at the Aral Sea proving ground. She died
of glanders, a disease that normally strikes down horses, in 1942.

Nothing else about her is known.

Dozens, perhaps hundreds, of people have given their lives in
the service of our scientific research. Occasionally their names crop
up in classified files, but many deaths have gone unrecorded. The
headstone on Rebirth Island was a rare public acknowledgment,
for those who possessed the security clearances to see it, of the
human cost of our program.

The history of Biopreparat and of the Soviet biological war machine is written in casualty reports, government decrees, instruction manuals for the manufacture of our pathogenic weapons, and
summaries of test trials. When I became second-in-command of
Biopreparat, I was able to obtain access to many of these records,
kept off limits to most employees. But even the records didn't tell
the entire story.

Over time, carefully, so as not to draw attention, I managed to
find out more through conversations with old-timers who remembered what the records did not, or would not, say. This is how I
learned that the Soviet Union's involvement with biological warfare began long before World War II.

A year after taking power in 1917, the Bolshevik government
plunged into a savage conflict with anti-Communist forces determined to bring down the fledgling workers' state. Red and White
armies clashed from Siberia to the Crimean Peninsula, and by the
time hostilities ended in 1921, as many as ten million people had
lost their lives. The majority of the deaths did not result from in-
juries on the battlefield. They were caused by famine and disease.

The casualties inflicted by a brutal epidemic of typhus from
1918 to 1921 made a deep impression on the commanders of the
Red Army. Even if they knew nothing of the history of biological
warfare, they could recognize that disease had served as a more potent weapon than bullets or artillery shells.

Victory in the civil war did not relieve the pressure on the new
government. Hostile foreign powers menaced the Bolshevik experiment on every side, and the weakened Soviet state seemed unlikely

to survive another onslaught. Someone realized that one of Russia's natural resources, its scientific talent, might help the revolution survive.

In 1928, the governing Revolutionary Military Council signed a secret decree ordering the transformation of typhus into a battlefield weapon. Three years earlier the fledgling Soviet government had signed an international treaty in Geneva banning the use of poison gas and bacteriological weapons. The weapons program was placed under the control of the GPU (the State Political Directorate), one of the predecessors of the KGB. It would continue to be supervised by state security organs until the early 1950s.

The 1928 decree represented a momentous decision. Bred in the unsanitary conditions of the battlefront or the slum, epidemic typhus has ravaged mankind for centuries. It is carried by lice from one infected person to another and cannot reproduce outside its host. Unlike typhoid fever, which is caused by salmonella bacteria, typhus is a rickettsial disease, carried by tiny rod-shaped microorganisms.

Once inside the body, the rickettsiae swarm through the blood, breaking down the cell walls of blood vessels as they multiply. Around seven to ten days after infection, victims will abruptly develop the first symptoms, beginning with throbbing headaches and a high fever. The stricken tissues become inflamed as they try to fight off the invaders, triggering a rash that spreads over the body. Spots of gangrene will sometimes appear on fingertips and other extremities as blood circulation slows down. Without treatment, the disease will send its victims into weeks of delirium and is fatal in 40 percent of cases.

Improvements in hygiene eradicated epidemic typhus from most of western Europe in the twentieth century, but it continues to afflict Africa, parts of South America, and Asia. A typhus vaccine was developed during World War II, though it is rarely used today, other than to immunize travelers to regions where the disease remains endemic. Administered in three separate doses over the course of five months, it provides almost complete protection from the disease. It was at one time also used for treatment, but it has been replaced in that capacity by antibiotics.

When the Soviet Union first turned to typhus, there was no

known way to control or contain this relentlessly efficient killer. The question facing our scientists was how to harness that efficiency.

Infecting lice with typhus and spreading them among a target population was not practical. Eventually someone hit on the idea of breeding typhus in the labs and spraying it in an aerosol form from airplanes.

Early biological weapons work involved primitive methods. The pathogens were bred in chicken embryos or in live animals such as rats that were killed when the concentration of pathogens was highest and were liquefied in large blenders. The liquid was then poured into explosives.

I learned about the 1928 decree and the early typhus experiments from a set of old reports at the Ministry of Defense. Summaries of experiments and testing, they were purposefully short on detail. No one wanted to commit the full information to paper. I can only assume that the original records have long since been destroyed. I was able to piece the rest of the story together with the help of veterans in our program who had learned the facts in turn from older scientists.

The first Soviet facility used for biological warfare research was the Leningrad Military Academy. Small teams of military and GPU scientists began to explore ways of growing significant quantities of typhus rickettsiae. The first attempts to cultivate typhus in the lab employed chicken embryos. Thousands of chicken eggs were sent each week to the Leningrad Military Academy—at a time when most Soviet citizens were lucky to get one full meal a day. By the 1930s, the Leningrad Academy had produced powdered and liquid versions of typhus, for use in a primitive aerosol.

Despite the program's secrecy, the Soviet government could not resist providing a public hint of its achievements. Marshal Kliment Voroshilov, a civil war cavalry hero who was Stalin's commissar for defense, declared on February 22, 1938, that although the Soviet Union planned to uphold the Geneva Protocol outlawing biological weapons, "should our enemies employ such methods against us, then I can tell you that we are ready—quite ready—to employ them against an aggressor on his own soil."

The biological weapons program soon expanded to harness other diseases. The Leningrad Military Academy sent some of its scientists and equipment one hundred miles north to the White Sea, a barren Arctic expanse flecked with tiny islands used to house political prisoners. By the mid-1930s Solovetsky Island, one of the largest, was the second major site of the Soviet biological warfare program.

At Solovetsky, a Soviet prison which later became the hub of Stalin's "Gulag Archipelago" concentration-camp system, scientists worked with typhus, Q fever, glanders, and melioidosis (an incapacitating disease similar to glanders). Solovetsky's large laboratory compound was built by prison labor. Many of the prisoners may also have been involuntary participants in our earliest experiments with biological agents.

The summary reports compiled by the Ministry of Defense describe several dozen cases of melioidosis from that period. The material I saw was intentionally vague as to whether humans were involved, but the way the case reports were arranged—with nineteen in one group, eleven in another, and twelve in yet another—suggested an irregular pattern not usually associated with animal testing. And the symptoms described could only have been experienced by human subjects. There have been repeated allegations in the West about Soviet germ warfare experiments on humans, but I have seen no other reports to indicate that these took place after the 1930s.

The research exacted a grim toll on our scientists. One account of a test with plague in the late 1930s ended with a cryptic note: "This experiment was not finished due to the death of the researcher." Another from the same period reported that twenty workers had been infected with glanders during experiments. The report didn't say where these experiments took place, or whether the workers had died, but in the days before antibiotics, death on exposure was virtually certain.

The biological agents explored before World War II underline the Soviet Union's primary interest in developing battlefield weapons designed to incapacitate enemy troops. Although this objective was reversed after the tularemia outbreak among our soldiers at Stalingrad, the laboratories in Leningrad and on Solovetsky Island were considered so crucial to Soviet defenses that when Nazi tanks invaded Russia in 1941, the high command ordered the immediate evacuation of both sites.

The lab equipment, fermenters, and glass vials containing strains of diseases were loaded on a train and sent south to the city of Gorky. On the day they arrived, Germans subjected Gorky to its first—and only—aerial bombardment of the war. Panicked commanders ordered the train to keep moving.

They came to a stop at Kirov, west of the Ural Mountains. The commanders of the expedition expropriated an army hospital for the severely wounded on Oktyabrsky Prospekt in the center of town (the patients were sent elsewhere) and the equipment was hastily reassembled. A new production line was working within weeks. It soon proved its value to the war effort. The lieutenant colonel who told me about the tularemia production line at Kirov also suggested that an outbreak of Q fever among German troops on leave in Crimea in 1943 was the result of an attempt to use another one of the biological warfare agents developed by his facility. I was never able to investigate this further, but Q fever was practically unheard of in Russia prior to that outbreak.

The region was swamped with refugees, transplanted munitions factories, and relocated airplane assembly plants when the production team arrived in Kirov. Scientists worried that the loss of Solovetsky Island left them with no place to test their agents. They started a search for a new testing ground, safe from the Germans and remote enough to avoid infecting the civilian population. The search led them to Rebirth Island.

The Soviet Union's approach to biological warfare took a new turn in September 1945, when Soviet troops in Manchuria overran a Japanese military facility known as Water Purification Unit 731.

Unit 731 operated Japan's secret germ warfare program. Rumors of the unit's activities in northern China had been circulating in Russia and the West since the late 1930s, but the details finally emerged through captured documents and the testimony of Japanese prisoners of war. The unit, commanded by Lieutenant General Shiro Ishii, experimented with anthrax, dysentery, cholera, and plague on U.S., British, and Commonwealth POWs. During the

Japanese invasion of Manchuria, porcelain canisters of fleas infected with plague and other primitive biological weapons were used in air raids that killed thousands of rural Chinese.

The captured Japanese documents were sent to Moscow, where they made absorbing reading. They included blueprints for biological warfare assembly plants, far larger and more complex than our own. Japan's program had been organized like a small industry, with a central production facility fed by continuous research and development.

Stalin ordered his most trusted aide, the sadistic KGB chief Lavrenty Beria, to match and if possible surpass what the Japanese had accomplished. In 1946, a year after the war ended, a new army biological research complex was established at Sverdlovsk. Construction engineers followed the designs laid out in the captured Japanese blueprints.

Stalin died in 1953. Beria was executed the same year, after an abortive attempt to seize power in the Kremlin. Under the new Soviet leader, Nikita Khrushchev, the responsibility for biological warfare was transferred to the Fifteenth Directorate of the Red Army. Colonel General Yefim Smirnov, chief of army medical services during World War II, became commander.

Smirnov was an impassioned advocate of biological weapons. He believed that they would dominate the battlefield of the future. A physician who had served briefly under Stalin as minister of health, he transformed the program into a strategic arm of the military and remained a dominating presence in the Soviet biological warfare program for the next twenty years. Smirnov worked so swiftly that Defense Minister Marshal Georgi Zhukov could announce in 1956 that Moscow was capable of deploying biological as well as chemical weapons in the next war—an announcement that set off a flurry of new offensive research in the West. Few Soviet citizens were aware of it.

By the late 1950s, facilities investigating every aspect of biological warfare were dotted across the country.

One of the most successful programs was created by the Ministry of Agriculture. A special division was established to research and manufacture anti-livestock and anti-crop weapons. The division was given the uninspired title of Main Directorate for Scientific and Production Enterprises. The biowarfare program was code-named "Ecology."

Scientists at the agriculture ministry developed variants of foot-and-mouth disease and rinderpest for use against cows, African swine fever for pigs, and ornithosis and psittacosis to strike down chickens. Like anti-personnel biological weapons, these agents were designed to be sprayed from tanks attached to Ilyushin bombers and flown low over a target area along a straight line for hundreds of miles.

This "line source" method of dissemination could cover large stretches of farmland. Even if only a few animals were successfully infected, the contagious nature of the organisms ensured that the disease would wipe out agricultural activity over a wide area in a matter of months.

Many of the ministry's facilities were installed in the centers of towns and cities, to keep their military connection camouflaged. This suggests how little those who ruled our lives worried about our health.

Across the street from the apartment block where I grew up in Alma-Ata (now Almaty), the former capital of the Soviet Socialist Republic of Kazakhstan, a large, rusting factory served as a makeshift playground for children in the neighborhood. It was a fantastic world of hulking machinery and cavernous tunnels, made all the more alluring by the large Keep Out signs posted conspicuously on the property. We would crawl through the fence on afternoons after school and, shifting through piles of metal, would occasionally stumble on odd-smelling canisters, painted in army green. Luckily, we never managed to open them.

Many years later, going through some old reports, I discovered that the factory was used by the Ministry of Agriculture until the early 1960s to make anti-crop and anti-livestock agents. It was called Biokombinat.

4

THE ENZYME PROJECT

Moscow, 1973

In the early days of the Cold War, when we appeared to be leading the world in space and nuclear weapons technology, Soviet biology was paralyzed. We had gone from being one of the world's powerhouses of immunological and epidemiological research to a backwater of demoralized and discredited scientists. The cause was one man—a Russian agronomist named Trofim Lysenko.

Lysenko came to national attention in the late 1920s, when he reported a successful experiment breeding winter peas in a remote farm station in Azerbaijan. His cultivation of several generations of plants resistant to cold temperatures led him to conclude that genetic theories about humans were wrong: rather than being a slave of his genes, man was capable of changing his essential traits through exposure to different environmental conditions.

Lysenko, who once bragged that he never reported the results of an experiment that contradicted his theories, claimed his work proved that environment was more important than heredity in the evolution of plants and animals. Calling genetics a bourgeois discipline that insulted the proletariat, he emerged as a paragon of the

"new" Soviet science based on Marxist materialism. By the 1940s
Lysenko was a confidant of Stalin. With the patronage of the Soviet dictator, he maneuvered his way to the top of the Soviet scientific establishment, imposing, in the process, an iron brand of
political correctness on the nation's biologists.

Dissenting scientists were condemned to prison camps or publicly humiliated. No journal that published an article on genetics
could survive. By the 1950s, little remained of the pioneering spirit
of Russia's great biologists and geneticists.

The gap in our scientific knowledge was of no interest to the
strategic planners responsible for modernizing our weapons program after World War II. Genetics didn't seem to have any connection to biological warfare. But a series of brilliant discoveries
between the 1950s and 1970s unleashed a revolution in Western
science, forcing the Soviet Union to recognize that it was behind in
more ways than one.

In 1953, two young scientists, James Watson and Francis Crick,
identified the shape of DNA, the genetic code that determines the
behavior of all life on earth. Over the next two decades, researchers found ways to manipulate DNA in the laboratory. They
discovered that genes of separate organisms could be cloned and
spliced together, a process that opened a new frontier in the study
of the behavior and treatment of disease.

Soviet biologists knew about the work in the West thanks to
smuggled journals and reports, but research conducted in Russian
labs was heavily restricted. The influence of Lysenko—who lived
until 1976—was too powerful. A few experts recognized that the
ability to manipulate genes broadened the horizon of bioweaponeering, offering the possibility of producing new strains capable of
overcoming vaccines and antidotes. To some, it also raised the disconcerting possibility that our competitors in the West could put us
at a severe strategic disadvantage.

Only one scientist had the clout, and the courage, to speak up.
His name was Yury Ovchinnikov, vice president of the Soviet
Academy of Sciences and a renowned molecular biologist.

Ovchinnikov understood the significance of what he had read
in Western scientific journals, and he knew that there were no Soviet laboratories, and few Soviet scientists, equipped to match that

level of work. He decided to resolve the crisis in Russian biology by appealing to the self-interest of the masters of our militarized economy. In 1972, he asked the Ministry of Defense to support a genetics program devoted to developing new agents for biological warfare.

Our practical-minded generals, like their counterparts everywhere in the world, were conservative and not easy to convince. Few of them knew the extent to which the Soviet Union was already committed to biological warfare, and even those who understood the concept had become skeptical of the extravagant claims for their importance made by old warhorses like Smirnov and Zhukov. They wanted weapons that would fire, explode, blast—not germs that no one could see. But Ovchinnikov was persuasive. The most skeptical military commander would have to agree that it was dangerous, if not outrageous, to be behind the West in anything.

Ovchinnikov found an influential ally in Leonid Brezhnev. The one-time metallurgical engineer who led the Soviet Union for eighteen years until his death in 1982 regarded the magisterial *akademiks* of the Soviet scientific establishment with a respect bordering on awe. Ovchinnikov was soon giving private lectures on genetics to Brezhnev and his aides. Slowly, the message sank in. Ovchinnikov, the youngest academician in the country, was appointed to a state commission exploring the military implications of the new gene-splicing technology.

The commission's work led to the most ambitious Soviet arms program since the development of the hydrogen bomb. Launched by a secret Brezhnev decree in 1973, the program aimed to modernize existing biological weapons and to develop genetically altered pathogens, resistant to antibiotics and vaccines, which could be turned into powerful weapons for use in intercontinental warfare. The program was called Enzyme.

The 1973 decree led that same year to the founding of Biopreparat. The nation's best biologists, epidemiologists, and biochemists were recruited in an effort that would soon absorb billions of rubles from the state treasury and spawn the most advanced program for genetically engineered weapons in the world.

The Enzyme project focused on tularemia, plague, anthrax, and glanders—all diseases that had been successfully weaponized by our military scientists but whose effects had been undermined by the development of antibiotics. But there were many other agents under review, including viral agents such as smallpox, Marburg, Ebola, Machupo, Junin, and VEE.

The Soviet Union's biowarfare research was concentrated at army factories in the cities of Sverdlovsk, Kirov, and Zagorsk. These were the only sites classified as "hot mode"—sufficiently insulated for work with highly infectious organisms.

Over the next decade, dozens of biological warfare installations disguised as centers of pharmaceutical or medical research were built throughout the country. In Leningrad, the Institute of Ultra-Pure Biopreparations was created to develop new techniques and equipment for cultivating pathogenic agents. At Omutninsk, in the pine forests near Kirov, a bacteriological research and weapons production facility was constructed alongside an old munitions plant operated by the Ministry of Defense. An entire "research city" for genetic engineering went up at Obolensk, just south of Moscow, and the Lyubuchany Institute of Immunology was established in Chekhov, also in the Moscow region, to investigate antibiotic-resistant disease strains. For work on viruses, the enormous Vector research and testing compound was built near the Siberian city of Novosibirsk.

These were just some of the facilities opened by Biopreparat. Existing state laboratories and research centers were also sucked into the new world created by Brezhnev's program. Several biological facilities managed by the Ministry of Health, including the large anti-plague research complexes in Kuybyshev, Minsk, Saratov, Irkutsk, Volgograd, and Almaty, were given special funding for weapons-oriented genetic research. The Soviet Academy of Sciences also played a significant role, conscripting four Moscow-region institutes into the Enzyme project: the Institute of Protein, the Institute of Molecular Biology, the Institute of Biochemistry and Physiology of Microorganisms, and the Institute of Bioorganic Chemistry.

Meanwhile, our testing program accelerated. Between 1979 and 1989, the Soviet Union conducted large-scale tests of an aerosol containing *Bacillus thuringiensis*—a harmless simulant—over the Novosibirsk region, using a plane with civilian markings. Similar experiments were run at a military proving ground near the city of Nukus in the Kara Kalpak Republic, and in the Caucasus. Another harmless agent, *Serratia marcescens,* was used in several tests conducted by the Institute of Biological Machinery inside the Moscow Metro system during the 1980s. Ballistic missiles containing simulants of biological agents were fired in tests over the Pacific Ocean between 1960 and 1980.

To manage the vast outlay of funds, a special department was created within Gosplan, the state economic planning committee. The operating and capital budget, considered too secret to keep in the hands of the civilian apparatchiks who ran every other sector of the Soviet economy, was administered by a high-ranking general.

Our program paralleled the Soviet nuclear complex in organization and secrecy. Both generated a sprawl of clandestine cities, manufacturing plants, and research centers across the Soviet Union. The atomic weapons network controlled by the Ministry of Medium Machine Building was much larger, but the production of microbes doesn't require uranium mines or a massive work force. When our biological warfare program was operating at its peak level, in the late 1980s, more than sixty thousand people were engaged in research, testing, production, and equipment design throughout the country. This included some thirty thousand Biopreparat employees.

Money was never a problem. As late as 1990, when Soviet leader Mikhail Gorbachev was promising the world major cutbacks in our arsenals, I was authorized to spend the equivalent of $200 million, including $70 million for new buildings. The total figure spent that year on biological weapons development was close to a billion dollars.

Biopreparat was the "brains" of the weapons program, supplying the scientific and engineering expertise for the projects commissioned by the army command. A special council, the Inter-Agency Scientific and Technical Council, acted as an advisory board. Headed by a government minister, the commission comprised twenty-five members from the principal scientific organizations of the country. Before I left Biopreparat in 1992, I served as deputy chief, with Kalinin. The chairman was Valery Bykov, then minister of medical industry.

Yury Ovchinnikov lived long enough to see his original ideas bear fruit. He died of cancer in 1987, when still in his fifties. I saw Ovchinnikov only once, at a large meeting in the Biopreparat offices. He was tall, charismatic, and elegant, very much like Kalinin. The two men knew each other well, and Ovchinnikov's quiet patronage was probably the deciding factor in transferring Kalinin, then an ambitious and relatively unknown officer in the Army Chemical Corps, into the coveted new agency when it was established in 1973.

Ovchinnikov had rescued Soviet biology from the morass of ideological politics, only to harness it to Soviet militarism. Although his name now graces a prominent Moscow science institute, he is remembered by many of us as the father of our modern biological warfare program. As Ovchinnikov recognized, such a program can only be as good as its scientists. The challenge was to find scientists willing to lead secret lives.

In April 1975, two months before I graduated from the Tomsk Medical Institute, a polite white-haired man in civilian clothes came from Moscow to the drab industrial town in Siberia where I had spent the previous two years in graduate study.

He wanted to meet several of the students who had specialized in epidemiology and infectious diseases. By then, I was one of them. In the intervening years, I had attended lectures on all forms of weapons of mass destruction and learned methods of protecting troops against nuclear, biological, and chemical attacks. No one ever suggested that we had a biological weapons program of our own. Instead, we were warned that as our enemies had them, it was vital for us to understand how they worked.

As I had learned from my brief foray into the Battle of Stalingrad, biological warfare was not the sort of thing you discussed openly. But I was fascinated by this area of military medicine. The romantic image of medics saving lives amidst the smoke and drama of a battlefield had appealed to me since I was a young boy. It struck me that military physicians were soldiers after their own fashion, waging a private war against an enemy that knew how to exploit every human weakness. The only weapons available to us were our skills in identifying symptoms and in applying the correct treatment.

My interests in epidemiology and in laboratory research were an ideal combination for the secret agency created two years earlier. This enthusiasm impressed my teachers. Aksyonenko must have passed my name on to the mysterious government recruiter, along with other students who showed the same passion for exploring the behavior of diseases.

Our visitor was soft-spoken and courteous. We were impressed that he had been given a special office to meet each one of us privately. I eventually learned he was a colonel from the human resources division of Biopreparat. He died several months after our meeting in Tomsk. I will never forget that meeting.

He was dressed in a dark suit and tie, but he carried himself erect, like a military man. He shook my hand with a firm grip.

"You have a good record and excellent recommendations from your teachers," he began. "Do you enjoy research?"

"Yes, sir," I said at once.

"Good," he smiled. "You seem to be exactly the sort of person we're interested in."

"What for?" I asked.

"I work for an organization attached to the Council of Ministers," he said elliptically, "and we could use your skills. I'm not permitted to say more, but I can tell you it has something to do with biological defense."

When I heard the phrase "Council of Ministers," I was thrilled. It was the highest government body in the land, redolent of power and authority. The prospect of working in a secret program for the state excited me, as did the idea of living in Moscow—which I automatically assumed was part of his invitation.

I also assumed he was not telling the entire truth when he spoke of biological defense. A special knowledge comes with growing up in a state like the Soviet Union. You were constantly alert to the probability that what you were being told had little relation to the message that was being conveyed. But the fact of the matter was that, at the age of twenty-five, I was too flattered by his attention to care.

"I'm interested," I said.

"Naturally," he went on, watching me carefully, "nothing can be final until we check you out. I'm going to give you a number of forms. Answer every question, in detail, and bring them back to me."

I stood up, forms in hand, and turned to go.

"One more thing," he called out. "Don't tell your friends or teachers about this conversation. Not even your parents."

The interview had lasted less than ten minutes, but it was enough to inspire in me a sense of the significance of what I was being asked to do. I obeyed his order almost to the letter. I called my parents and told them I might be getting an important assignment in Moscow, but that they would have to wait before I could tell them more.

A few weeks after the interview, we stood in our crisp new junior lieutenants' uniforms and polished knee-length boots on the parade ground of the institute. It was graduation day, and the commanding officer began to read out the names of every newly commissioned lieutenant and his assignment. A few students received coveted postings in East Germany or Poland. Others were condemned to the backwoods boredom of a provincial army base.

My name wasn't called until the end of the ceremony.

"Lieutenant Kanatjan Alibekov!"

I stepped forward and saluted.

"You are assigned to the Council of Ministers of the Soviet Union!"

The names of four other classmates followed, all with the same assignment. Unbeknownst to me, they had also been interviewed by the friendly white-haired man.

I couldn't help but grin: I was going to Moscow.

Several days later, each of us was called to the school's administrative office to receive our letter of assignment. I looked it over quickly, and my face fell.

I was assigned to a post office box.

"What does this mean?" I asked. "Where is this?"

The officer who gave me the letter tried not to smile when he saw my expression.

"Omutninsk," he said. "It's near Kirov, but you're not supposed to tell anyone. You'll be getting a letter of authority, which you can use to get a train ticket."

All five of us, it turned out, were going to the same place. Some of our other classmates had been impressed when they first heard of our assignment to the Council of Ministers. A few shrewder students understood that we were pointed toward secret work. Some even guessed that we were going to biological "research facilities," although no one was quite sure what they were, and no one dared ask.

"You're going to have a very short life," one of my friends said breezily. "I've heard no one lasts in those programs more than a couple of years."

BIOHAZARD

5

LAB WORK

Omutninsk, 1975

I have lost all sense of smell and have the broadest range of allergies of anyone I know. I can't eat butter, cheese, eggs, mayonnaise, sausages, chocolate, or candy. I swallow two or three pills of anti-allergy medicine a day—more on bad days, when my sinuses start to drain. Every morning, I rub ointment over my face, neck, and hands to give my skin the natural lubricants it has lost. The countless vaccinations I received against anthrax, plague, and tularemia weakened my resistance to disease and probably shortened my life.

A bioweapons lab leaves its mark on a person forever. But this was all in the distant future when I stepped off the train in a desolate corner of western Russia on a wet midsummer's night in 1975.

The East European Scientific Branch of the Institute of Applied Biochemistry was tucked away in a forest carpeted with mushrooms just outside the old Russian city of Omutninsk. It was almost a city of its own. More than ten thousand people lived and worked there, nearly a third of the population of the nearby town. Some thirty weatherbeaten brick buildings, including dormitories, labs, schools, and a heating plant, dotted the grounds. The working area was surrounded by a concrete wall and an electric fence, but the entire complex could have been mistaken for any of the self-sufficient civilian industrial enterprises built by the dozens in equally remote areas of the country. Trucks lumbered in and out every day. Schoolchildren played in one section of the compound. The guards at the front gate never wore uniforms.

Omutninsk housed one of Russia's newest biological warfare facilities. A chemical plant that had been producing biopesticides in the compound since the 1960s was expanded by the Fifteenth Directorate to serve as a reserve "mobilization" plant for wartime production of biological weapons. In the 1970s, construction began on a new complex of buildings. When I arrived, two years after Brezhnev's secret decree, Biopreparat was in the process of turning Omutninsk into a major center of biological weapons production.

In organizational charts the compound was designated as the "Omutninsk Scientific and Production Base," but we referred to it in our coded cables by its post office box number: B-8389. Officially, Omutninsk manufactured pesticides and other agricultural chemicals. Unofficially it served as a training ground for the next generation of Soviet bioweaponeers.

Some ten or fifteen of us, all in our twenties, arrived that summer. Freshly commissioned officers, we came from military graduate institutes around the Soviet Union. Several had been trained, as I had, in medicine, but our group also included engineers, chemists, and biologists—picked after mysterious interviews followed by long background checks to ascertain that no hint of subversion lurked in our families.

From the very first night when I arrived, soaking wet, to report to my new commanding officer only to be chastised for wearing my military uniform, I knew I had entered into a new world. There were no orientation lectures or seminars, but if we had any doubts as to the real purpose of our assignment, they were quickly dispelled. We were given a paper with a list of regulations for behavior at the plant. At the bottom we were made to sign off on a pledge never to reveal what we were told or what we did.

Our "instructors" came from the KGB. They handed us more forms explaining that we would be doing top-secret research in biotechnology and biochemistry for defensive purposes. Then we were called, one by one, for individual sessions.

"You are aware that this isn't normal work," the officer told me as I sat down. It was a declaration, not a question.

"Yes," I replied.

"I have to inform you that there exists an international treaty on biological warfare, which the Soviet Union has signed," he went on. "According to that treaty no one is allowed to make biological weapons. But the United States signed it too, and we believe that the Americans are lying."

I told him, earnestly, that I believed it too. We had been taught as schoolchildren and it was drummed into us as young military officers that the capitalist world was united in only one aim: to destroy the Soviet Union. It was not difficult for me to believe that the United States would use any conceivable weapon against us, and that our own survival depended on matching their duplicity.

"Good," he said with a satisfied nod. "You can go now—and good luck."

The five minutes I spent with him represented the first and last time any official would bring up a question of ethics for the rest of my career.

Bacteria are cultivated identically whether they are intended for industrial application, weaponization, or vaccination. Working first with harmless microorganisms, we were taught how to make nutrient media, the broths in which they multiply. Making these potions is an art in itself. Bacteria require highly specialized mixtures of proteins, carbohydrates, and salts—often culled from plant or animal extracts—to achieve the most efficient growth rate.

We would take samples of the nutrient media and analyze their biochemical components, testing for pH and amino acids and calculating the concentration of carbohydrates and other compounds. Then we mixed in the seed material—the bacterial agent—to determine its quality, concentration, and viability. The process of seeding the agents was a delicate one and had to be performed under perfectly aseptic conditions. Next, we studied how temper-ature, oxygen concentration, differing components of nutrient media, and countless other factors affected bacterial growth.

Within months, I would move from the simple lab techniques of medical school to complicated industrial procedures in bio-chemistry and microbiology. For the first time in my life, I would work with pathogenic agents, learning how to infect lab animals and conduct autopsies.

Russian bioweaponeers divide their facilities into three zones, rated according to the safety of the materials with which they work. (Most countries have four.) Zone One is restricted to the preparation of nutrient media. Zones Two and Three are both "hot zones," sealed off from the outside world with special filtra-tion systems. Zone Three at Omutninsk throbbed with the con-stant hum of steel dryers and centrifuges. In this zone, we had to wear bubble helmets, large gloves, and thick rubber outfits which we called "space suits." They gave us the slow, tentative stride of astronauts walking on the moon.

Zone Two had its own protective gear—not quite so cumber-some as the space suit, but still requiring an elaborate rite of pas-sage from the outside world. To enter Zone Two we would shed our white lab coats and pants and put on a long surgeon's smock, stretching down to the ankles, and a cloth hood with openings for the eyes and nose. Over the hood we placed a sealed respirator mask. Then came high rubber boots and a pair of thin rubber gloves—two pairs if we were going to work with animals.

My first weeks at Omutninsk were exciting—and excruciating. I had been inside laboratories in medical school, but I had never seen a lab as large and forbidding as the one we were brought into on the first day. White tables stretched from one end of the room to the other, topped with microscopes, photometers, and row upon row of glistening glass test tubes and flasks.

We were given white lab coats, divided into smaller groups, and assigned to a lab technician who would serve as our mentor. My first trainer was a young woman named Svetlana, a blonde with blue eyes and a perpetually amused expression as she led us through our lab work. I was half in love with her, which made it all the more embarrassing when the fragile flasks she ordered me to sterilize kept shattering in my hands.

"What a bear this one is!" she said to one of my lab partners.

I thought I'd never get the hang of it.

But slowly I gained confidence in a laboratory world that seemed to offer new discoveries every day. I learned how to use the delicate pipettes to transfer liquids from one vial to another as we heated them over Bunsen burners. The magic of making cultures grow from minuscule particles barely discernible under the microscope fascinated me.

We would arrive at the lab at eight o'clock in the morning, breaking only for lunch in the small cafeteria at noon before returning to our microscopes and test tubes to work until dinner. Sometimes we spent the afternoon in the library poring over scientific texts, which we were expected to summarize at weekly conferences. The lab training was only a prelude to our initiation into the heart of Omutninsk's mysteries: the giant reactors in which the industrial production of pesticides took place.

The workers inside the industrial building were less tolerant of our youthful mistakes. They knew nothing of the real purpose of our training: as far as they were concerned, we were "fancy" university grads who had never gotten our hands dirty. We were set to work scrubbing floors and washing machines until we were considered trustworthy enough to assist in the running of the plant. The microorganisms we cultured inside the huge vats were harmless—such as *Bacillus thuringiensis*—but the complicated procedures involved in manufacturing them represented a rehearsal for our future work with pathogenic agents. The liquid cultures were transported between different buildings in the complex through special overhead pipes. The most important element of the process was to ensure that the cultures stayed pure from beginning to end. The need to keep our working materials and our machines sterile was drummed into us every minute of the day. Too tired and too excited to sleep when we returned to our dorms, we found relief from the pressures of the labs by venturing into town at night.

Omutninsk was a placid community of single-family wooden houses and narrow streets whose inhabitants seemed pleasantly oblivious to our activities. They had long experience in the art of weapon making. In the seventeenth-century, Peter the Great built an iron foundry in Omutninsk that became one of Russia's first weapons factories, turning out primitive cannons for the czar's armies. Three centuries later, military production still dominated the local economy. A dilapidated metallurgical plant responsible for supplying parts for rifles and artillery pieces employed most of the townspeople.

If the people of Omutninsk seemed to have little interest in what was going on in their backyard, some of us were beginning to wonder what we had gotten ourselves into.

Over quiet drinks in the town's only restaurant we argued incessantly about the work we were being groomed for. Some of the young scientists felt proud to be associated with secret affairs of state. Others were repelled by the idea of turning diseases into weapons—even if the project was defined as a national priority.

One of the four medical graduates who came with me from Tomsk, a tall, burly Siberian named Vladimir Rumyantsev, grew petulant and increasingly depressed. After returning from the restaurant, he would lie on his bed and stare at the ceiling for hours, nursing a bottle of vodka. We had become close during training, and I felt freer to exchange confidences with him than with others in the group.

"Kan, we're doctors!" he once exclaimed. "How can we do this?"

I was asking myself the same question. In the Oath of a Soviet Physician, which I'd taken at graduation, I had pledged to help the sick, "to do no harm," and to be on call "day, night, and vacation period." So far, I was principally fulfilling the third part of my pledge.

But I liked the lab work. I discovered an affinity for the meticulous processes involved in culturing organisms. The challenge of manipulating the tiny worlds that appeared under my microscope engaged me more intensely than anything I had ever done before. In the evenings, in medical texts and journals borrowed from the library, I read about the behavior of diseases until each one took on a distinct personality in my mind. I knew that the results of my studies could be used to kill people, but I couldn't figure out how

to reconcile this knowledge with the pleasure I derived from research.

About four months into training, I decided to escape. It wasn't a
bold attempt—I didn't want to confront the KGB or my military
superiors, who seemed to have such a high sense of my potential—
and I wasn't too surprised when it failed.

I took the train to Kirov, five hours away, and mailed a long letter to my father. I had not told my parents about my assignment,
except to say that it was related to secret military matters. I
thought I could avoid KGB interception by posting the letter as far
as possible from the base, I still kept my words as vague as possible. I assumed my father could read between the lines. He knew the
military far better than I.

My father, Bayzak Alibekov, was wounded seven times in
World War II. He was decorated for bravery in the historic tank
battles around Kursk and had risen to the rank of lieutenant
colonel of police in Alma-Ata, after four frustrating years as a rural
policeman. Our family boasted an illustrious lineage: my grandfather had been a hero on the Communist side of the Russian civil
war. He was Kazakhstan's First People's Commissar of Internal Affairs, responsible for police and security in the 1920s and 1930s,
and a street in Kazakhstan's old capital was named after him. Perhaps my background would give me an honorable way out.

In the letter, I asked my father to write to Marshal Andrei
Grechko, then the minister of defense and one of the country's
most admired military figures, to request that I be reassigned.

A week later I called my father from the long-distance telephone office in Omutninsk.

"Are you sure you want to do this?" he asked.

"Yes," I said eagerly. "You could tell the marshal that, as a
wounded war veteran, you need your son close to home."

"That's true enough," he laughed. "I'm becoming a deaf old
man."

He didn't ask me what I was doing for the army, and I didn't
volunteer any information. He protested at first, but when he
heard the anguish in my voice, he agreed to write the letter.

A warm and respectful note addressed to my father arrived in

Alma-Ata a month later. "Dear Comrade Alibekov!" it began. "I salute you for your services to the Motherland, and I respect your wishes regarding your son. However, you must know that your son has been chosen to conduct very important work for our country, and we cannot spare him. It is always important for a son to be with his elderly parents, but of course you have another son and daughter living close by who can perform their obligations."

My father read Grechko's words to me over the telephone. He couldn't contain his excitement: he had received a letter signed by the great marshal himself! Meanwhile, I felt trapped.

Not long after that I began to feel proud—and a little flattered. If the defense minister himself regarded me as irreplaceable, who was I to argue with him? I could learn to enjoy being part of this strange and secret club.

I applied myself to the training with more energy than I had shown in weeks and gave up, for the time being, all thoughts of leaving. In the end, no one in our group left the program.

New buildings were going up all around the compound. Every day prisoners from a nearby labor camp were bused in with shovels and cement mixers. Biopreparat had signed a secret contract with the Ministry of Internal Affairs to employ prisoners doing hard time of ten years or longer as the principal construction force for new biological facilities around the country.

The foundations were laid for a pilot plant that would usher in a new era in the mass production of biological weapons, a plant that came to be known as Building 107. The flurry of construction activity left little doubt that we were part of a crucial national defense drive.

I soon had other reasons to value the status that came with my career. When I was transferred to a new post in Siberia, in March 1976, I had met the woman who would become my wife.

Lena Yemesheva was an attractive eighteen-year-old foreign-language student when we met in Alma-Ata earlier that winter. I was home for my first leave since going to Omutninsk. She was a friend of a cousin, who brought her to a Soviet Army Day concert in town. I liked her immediately. She had shining green eyes and came from a town close to the settlement in southern Kazakhstan where I had been born. Lena had studied physics before she was transferred to the Alma-Ata Institute of Foreign Languages, so we had in common a love of science as well.

We were married in August 1976 at the wedding palace in Alma-Ata and went to her father's home for a *toi*—the traditional Kazakh wedding feast.

Midway through our courtship, Lena had stopped asking me questions about my job. She had grown tired of my enigmatic reply that I was engaged in "secret work." Still, I was surprised at how readily she accepted the news that we would be making our first home together in the faraway Siberian city of Berdsk.

As she explained much later, all that counted in those days was the satisfaction of knowing that I was a military officer clearly marked out for the favors of the state.

The Berdsk Scientific and Production Base, or the Siberian Branch of the Institute of Applied Biochemistry, as it was formally called, was a welcome change from Omutninsk. Berdsk, about twenty-five hundred miles from Moscow, formed part of one of the most impressive scientific communities in the country. It was near the "academic city" of Novosibirsk, a center for advanced research in engineering, technology, and economics. Mikhail Gorbachev would later recruit his first *perestroika* brain trust from Novosibirsk's gifted group of economists and political scientists. The most interesting industrial landmark in Berdsk was a radio assembly plant built with the help of U.S. technicians in the 1940s. My assignment there soon made me forget my crisis of conscience the previous summer.

Biopreparat wanted to transform Berdsk into a prototype of a combined research center and weapons assembly line. The facility, built in the 1960s, had been used mainly as a reserve plant for assembling and filling bomblets with weaponized bacteria that would be produced at its own and other installations. As the Soviet bioweapons program expanded, such a division of labor proved cumbersome. Berdsk had no way of guaranteeing that the bomblets it produced wouldn't leak and threaten our own soldiers. Liquid or powdered simulants weren't adequate for leakage tests;

you needed the real thing. But Berdsk didn't have a research and development lab of its own. Moscow headquarters ordered that one be built and purchased equipment from overseas to stock it. Rumyantsev and I found ourselves assigned together to a crash program aimed at turning the facility into an installation capable of developing new production techniques and formulations.

Hundreds of unopened boxes filled with new machines were piled up inside the facility when we arrived. They were intended for a new microbiology lab, but they had been sitting around for months. The staff only knew how to run an industrial assembly line; they had no idea how to create a laboratory.

Rumyantsev and I built a microbiology lab from scratch. We planned the layout of the room from the sterile working tables to the sinks and water pipes. Gradually unpacking the boxes, we pulled out microscopes, test tubes, ovens, and a collection of equipment gathered from every part of the world. There were U.S. and Japanese fermenters, Czechoslovak reactors, French-made flasks. The bulk of our equipment came from the United States and Great Britain. We owed a lot more to our old allies than anyone could publicly concede. The fact that we could use standard fermenting machinery was a vivid illustration of the dual nature of the tools of our trade.

Within three or four months, we presented our managers with a fully equipped laboratory.

In January 1977, the commander of Berdsk, Colonel Vitaly Kundin, returned from a visit to Biopreparat headquarters carrying two small ampoules filled with freeze-dried *Brucella*, a bacterium that preys on cattle and other livestock. Transmitted to humans as brucellosis, or Malta fever, the bacteria cause fever, sweats, a sore throat, and a dry cough, sometimes accompanied by vomiting, acute abdominal pains, and diarrhea. Even when treated, the illness can last for months, and it can become chronic. My knowledge of this disease came from other sources besides textbooks. My father suffered for years from recurring bouts of brucellosis that left him in so much pain he was sometimes unable to move his hands.

"Now that we have a laboratory, we have something to use it

for," Kundin said cheerfully. "Why don't you fellows see what you can do with this?"

None of the Biopreparat labs manufactured brucellosis in any significant quantity. The standard growth medium used until then contained a milk protein called casein. Examining some of my textbooks, I discovered a recipe for a mixture of yeast extract, vitamins, and other growth stimulators that had produced high yields with other cultures. In our new lab, Rumyantsev and I spent long hours experimenting with various combinations of the mixture until we got it right.

After eight months of work, we presented our findings: our new growth medium produced a substantial yield of bacteria that could be weaponized. Moscow headquarters was pleased. For me, it was a personal achievement—I had gone from being a pupil to a practitioner.

In the fall of 1977 I was promoted to senior lieutenant and senior scientist for my work in Berdsk. I also became the head of a new family: Lena gave birth to our first child, Mira, that year, and our lives seemed perfect. With my growing salary and the professional esteem of my peers, I began to believe that I had found the best of all possible worlds in the Soviet Union.

Two years later, when Lena was pregnant with our second child, Alan, I was named acting lab chief at Berdsk. After Alan was born I was promoted again and handed a new assignment: I was ordered to return to Omutninsk to develop a production process to weaponize tularemia. Building 107, they told me, was finally ready.

6

BUILDING 107

Omutninsk, 1980

Omutninsk was a hive of activity. There were new buildings everywhere in the compound, but the one that mattered most was the spotless gray three-story structure designated as the pilot plant for tularemia.

Building 107 was structured according to the box-within-a-box principle, to keep the deadliest organisms out of the surrounding countryside. If you could lift the roof, it would resemble a Russian matryoshka doll. Snuggled inside the matryoshka is another doll, and then a tinier one inside, continuing until you reach the limits of the craftsman's skill, or, in this case, workable space.

The outer shell of Building 107 was Zone One. It housed the offices of administrative and security personnel and laboratories used for noninfectious organisms. Workers dressed in plain white lab coats and white pants milled in the corridors. Daylight poured through the windows, and the walls were plastered with resolutely upbeat Party banners: "Fulfill our Five-Year Plan in four years!" "Long Live the Communist Party of the Soviet Union!" It was possible to spend an entire day in Zone One without being aware of what was happening deeper inside the building.

Zone Two contained "hot" laboratories for work with pathogenic materials, storage vaults, animal cages, and giant sixteen-ton and twenty-ton fermenters, which soared to the upper levels of the building. Zone Three, nestled inside Zone Two, displayed the fruits of our engineering efforts since 1973: rows of gleaming steel centrifuges and drying and milling machines.

Both inner zones had their own air supply system. Noisy generators pumped air through an overhead latticework of exposed pipes, keeping the atmosphere inside at pressures slightly lower than normal to prevent contaminated air from seeping into Zone One. Hydrogen peroxide was sprayed into the air from nozzles in the ceiling. The distinctive smell of that particular disinfectant will stay with me forever. It wasn't just its smell that made an impression: for the dozen or so years I worked inside the labs, my black hair was bleached a dirty blond.

It was a world of invisible perils. One false step, a fumble, an unthinking gesture, could unleash a nightmare. We all knew enough to fear the hazards of the two hot zones, but we were young and felt invincible. We saw ourselves as custodians of a mystery that no one else understood, warriors or high priests of a secret cult whose rituals could not be revealed.

Late one Sunday night in March 1983, the phone rang in our apartment. I reached over to pick it up, trying not to wake Lena.

It was Nazil, one of the lab chiefs on night duty at Building 107.

"You'd better get down here," he said tersely. "We've got a problem."

I dressed quickly in the dark and hurried to the compound.

Once inside I went directly to the corridor that led to Zone Two. The corridor, called the "sanitary passageway," was a warren of small sterile rooms linked by a series of connecting doors. We entered through a sealed door with a coded lock. The door could be opened with a latch or by turning a heavy wheel, like a submarine. The codes changed once a week.

I stripped my clothes off and stuffed them into one of the lockers lining the walls. Then I walked into a second room, where a young nurse sat behind a desk. I had a nodding acquaintance with her outside the lab, having seen her walking her large dog, and was embarrassed at first to appear before her naked. But she always maintained a businesslike air as she wordlessly stuck a thermometer under my armpit and examined every inch of my body, including my teeth and gums. Any sign of bleeding from a cut or bruise, even from a nick while shaving, was grounds for barring further entry.

The buzz of the ventilators grew louder as I passed through the next rooms, picking up the separate items of my anti-plague suit: white socks and long johns, hood and cotton smock, respirator, goggles, boots, and gloves. The entire procedure was reversed on the way out, although the gloves were always the last to come off. Even with long practice, I never managed to complete the process in under fifteen minutes. That night, I was faster than usual.

Nazil was waiting for me inside Zone Two.

As we walked together down the corridors, he told me what had happened. The air pressure in the pipeline feeding one of the tularemia rooms had begun to drop precipitously. A technician had been working there an hour or so before, but she had gone home. She may have forgotten to reset the valves.

Nazil was anxious to get back to work before his shift ended. It was 11:00 P.M. He brought me to the room where the drop in pressure had been reported and hesitated at the door.

"Don't worry," I said. "Go back to your lab. I'm sure I can handle this."

Mollified, he set off down the corridor. I opened the door and took a few steps inside. It was pitch black. I reached back, groping in the darkness for the light switch. When I finally hit the switch and looked down, I found I was standing in a puddle of liquid tularemia.

It was milky brown—the highest possible concentration. The puddle at my feet was only a few centimeters deep, but there was enough tularemia on the floor to infect the entire population of the Soviet Union.

I called for Nazil, frozen in place, and heard him rustling toward me down the hall.

I was only two feet or so from the doorway, but I was trapped. If I tried to back out I would bring the disease with me into the corridor—and, potentially, into the rest of the zone.

Keeping my voice as calm as possible, I told Nazil to bring disinfectant quickly—anything he could find. I reached my gloved hand behind me and grabbed the bottle of hydrogen peroxide he handed through the partly open door.

I poured the solution over my boots. He handed me more bottles as I moved backward, tiny step by tiny step, pouring all the time.

By the time I was out of the room, three military scientists working in other parts of the zone had rushed to the scene, alerted by the commotion. The change in air pressure must have caused the culture to escape through the filter system. I closed the door and told them to disinfect everything I had touched, as well as the room itself.

I went back through the sanitary passageway, eased off my boots and protective suit, took a disinfecting shower, and submitted myself to a quick checkup by the nurse. She assured me that I was fine.

Silently, I congratulated myself on my good fortune. I tried to imagine what might have happened if I had lost my footing on the slippery floor. Although tularemia isn't usually deadly, we were working with a far more virulent strain than any I would ever have been exposed to in nature.

When we regrouped in Zone One, I advised Nazil and the others to take the antibiotics we had on hand for emergencies.

I went to my office and called Savva Yermoshin, chief of the KGB detachment at Omutninsk. Savva would later work with me at Biopreparat headquarters in Moscow.

I had obviously pulled him from a deep slumber.

"Savva, I'm sorry to wake you," I said. "I just wanted to let you know a small amount of tularemia was released inside Building 107 tonight."

I didn't expect him to do anything, but regulations required us to inform the KGB about the slightest break in routine.

"Anybody hurt?" he said in a voice fogged with sleep.

"No, it's all under control," I continued cheerfully. "We've got it cleaned up. There's nothing for you to do."

I looked at my watch after hanging up. It was almost 2 A.M.. It was pointless to call Moscow at that hour. I decided to wait until morning and went home, tired and relieved.

"What was the emergency?" Lena asked me sleepily as I padded around in the dark of our bedroom.

"Nothing important," I told her. "Go back to sleep."

Around lunchtime the next day I got a call from an extremely upset Kalinin.

"I've been trying to find you all morning and they keep telling me you're in meetings," he yelled. "How can you sit around in meetings when your building is leaking tularemia into the ground?"

Yermoshin, it appeared, knew more about the regulations than I did. He was supposed to inform his superiors whenever an emergency occurred, and he had duly contacted the KGB director for the Kirov region as soon our brief conversation ended. It didn't occur to me that such a minor mishap would need to be relayed up the chain of command, but the KGB chief in Kirov had called his bosses in Moscow, who called Kalinin early that morning.

By then, the story had become hopelessly mangled. Whatever the sleep-addled Yermoshin had told his senior officer, it had been magnified into a disaster threatening the entire region.

I tried to calm Kalinin down, but he didn't believe me. He had absolute faith in the KGB.

"I'm sending someone out there on the first train tomorrow morning," he said, and hung up.

The next morning I went to the station to pick up General Lev Klyucherov, the head of Biopreparat's scientific directorate. He arrived looking as if he'd spent the entire journey stewing with rage.

"Whatever you're trying to hide," he said at once, his face reddening, "it's not going to work."

I asked him to come into my office and went over the entire incident step by step. Klyucherov softened slightly and seemed persuaded. After all, he could see for himself that no one had fallen ill.

No one, that is, but me.

Toward the end of Klyucherov's visit, my body started to shake. Chills, and a sudden wave of nausea, overcame me so quickly I wanted to bury my head in my arms.

It's a cold, I thought. I've been working too hard.

But it felt worse than any cold I'd ever had. I could feel my face burning with fever.

"What's happened to you?" Klyucherov asked in a tone that was now much friendlier. "You look like you're about to die."

I smiled weakly. "It's just a cold," I said. "I had a long night. I could do with some tea."

I went home as soon as the general left. There was no doubt in my mind as to what was wrong: tularemia begins with flu-like symptoms and it moves through the body quickly.

At home I went straight to my small library of medical textbooks and took down from the shelves every book on infectious diseases I could find. Antidotes were not my field of expertise. I tried to think through my next step.

If this ever got out, Klyucherov, Kalinin, and everyone else in Moscow would make my life more miserable than it was even now. They would accuse me of trying to hide the seriousness of the incident and would wonder what kind of scientist would forget in such a situation to take the proper antibiotics. I had told Nazil and the others to take antibiotics, but for some inexplicable reason I hadn't taken any myself.

I felt humiliated and confused. By the time I'd left Building 107 that night, I had been completely disinfected. I must have caught the disease in a matter of seconds, between leaving the sanitary passageway and entering the shower. But how? Then it came to me. I must have brushed my face while taking off my mask and hood. A hundred cells, an amount smaller than a speck of dust, would have been enough to infect me through an imperceptible cut or scratch.

I knew it was safe to stay at home: there was no danger that the infection in my body would spread to Lena and the children.

Tularemia can be inhaled or ingested or contracted through bites or scratches. It rarely passes directly from person to person, but it can be carried by fleas, ticks, rats, and other rodents and can enter the bloodstream through minor abrasions. The disease is marked by a sudden onset of fever and chills, often followed by an incapacitating headache. As soon as it enters the body, the bacteria will begin to multiply locally, gradually spreading to lymph nodes and distant organs, including the liver and spleen.

Even after successful antibiotics were developed in the 1940s, tularemia was considered an ideal weapon for the battlefield due to the speed with which it could overwhelm an opponent's medical resources, leaving hospitals and physicians unable to cope with a flood of patients in need of constant treatment.

If taken immediately, antibiotics can contain the spread of the disease and kill invading bacteria in a matter of days. The later the drugs are administered, the longer a victim will suffer. Particularly acute cases have been known to linger for months.

Tetracycline was thought to be the best antidote for tularemia, but I had no way of knowing how well it would work against the strain we had produced in our lab. In exceptional cases, certain highly virulent strains are capable of overcoming ordinary antibiotic treatment and can be fatal.

I called a friend's wife, a physician at the local hospital, and told her I needed tetracycline urgently. Under normal circumstances I would have required a prescription, but in a small town it was easy to cut corners.

"How much?" she asked, without registering the least surprise.

Calculating quickly, I asked for three times the customary dose. There were advantages to Soviet secrecy. I would have had a hard time getting that amount of tetracycline in the United States without a good explanation. I told her not to tell anyone.

I wanted a high-impact, crash dose. If it didn't work, I'd have to check myself into the hospital. Self-treatment had its limitations.

An hour later, my friend's wife showed up with the pills in a cavernous shopping bag, the kind Russian women carry around for emergencies. Lena answered the door. I sat in the armchair in the living room, too sick to move.

By the end of the day, my fever had begun to drop. I stayed home the next day, after calling in with a cold. By Wednesday or Thursday, three days after my exposure, I was better, although I continued to take high doses of tetracycline for the next ten days. I was able to return to work the following Monday.

When Lena asked me what had happened, I told her I'd had a mild infection after an accident with one of the substances in the lab. She knew nothing about tularemia, since I never shared with her the details of my work. She pretended to be calmed by my assurances that nothing was wrong, but when we finally left the Soviet Union she confessed how frightened she had been.

I had covered up the incident, but it was a powerful reminder of the consequences of our trade.

7

ACCIDENT AT SVERDLOVSK

Sverdlovsk, 1979

No nation would be so stupid as to locate a biological
warfare facility within an approachable distance
from a major population center.
—Raymond Zilinskas, U.S. clinical microbiologist,
in a 1980 report on the Sverdlovsk accident

Biopreparat was the darkest conspiracy of the cold war, a network so secret that its members could not be told what colleagues in other parts of the organization were doing, or where.
Yet even the most furtive networks are made of human beings.
Gossip, professional rivalry, and ordinary curiosity ensured that
we always knew a lot more than our leaders imagined.

Eventually, everyone found out about Sverdlovsk.

I first learned what happened there in a typically, and madden-
ingly, casual manner. It was June 1979, two months after the accident had occurred, and I was in Siberia, struggling through a
period when nothing seemed to go right. The few successes I had
achieved felt dull and unimportant and, despite the recognition I
was getting from Moscow, I was convinced my career was going
nowhere. I had managed to persuade myself that the world had
passed me by.

There was no one to complain to except Lena, until I found a
sympathetic ear in a colonel sent to Berdsk on a routine inspection
trip from headquarters.

His name was Oleg Pavlov. He was the kind of man who would drop official business at a moment's notice for a glass of vodka and a chat. One Friday, after a tedious discussion of our research budget, he asked me if there were any places around town one could go to enjoy, as he put it, "the real Siberia."

I told him our workers spent weekend afternoons swimming and picnicking with their families at a nearby river. On a weekday, there would be no one there.

"Wonderful!" he boomed. "Let's get some refreshments."

It was a lovely, warm summer day. The park by the riverbank was deserted and the birch trees shifted gently in the breeze. Pavlov tore off his clothes and jumped into the water with a yell. I followed him, watching with amusement as he splashed like a child in the ice-cold stream.

We clambered to the shore, dried ourselves, dressed, and unwrapped packages of hard-boiled eggs, sausages, bread, and onions. Pavlov brought out a bottle of clear vodka and two glasses. We sat in the shade, blissfully contemplating the world.

In Russia, a glass of vodka is an invitation to expose your soul. Without planning to, I began to pour out my frustrations.

"I can't get anything done here!" I said. "There are never enough scientists, and we never do important work anyway. I wish they would give us something serious to do."

Pavlov swallowed the contents of his glass in one gulp and set it down on the riverbank.

"Don't be an ass," he said.

I was too stunned to speak.

"Let me give you some advice," he went on. "Never wish for something too hard, because you just might get it."

I wondered if the vodka had made me more candid than I ought to have been, especially with someone from headquarters.

I tried to save face. "I should never drink in the afternoons," I said, hoping that he would smile and change the subject.

But Pavlov didn't smile.

"You know about Sverdlovsk, don't you?" he asked suddenly.

I considered what to say. Most of us knew, unofficially, about the army biological research facility in Sverdlovsk, in the eastern foothills of the Urals. It was built after the war, using specifications found in the Japanese germ warfare documents captured in Manchuria.

"Well, I know they're doing anthrax work," I responded. "Have they had some kind of achievement?"

He shook his head in irritation. "You haven't been told? There's been an accident."

"What kind of accident?"

He poured out another glass of vodka, drank it down, and smiled mischievously.

"You're too young to hear about this kind of thing."

I begged him to tell me more, but he refused.

"I can't tell you if you don't already know," he said in an exasperated voice. "I only brought it up to show you how lucky you are not to be doing the kind of work you want to do, the 'important work' they were doing in Sverdlovsk. You're young, you're happy, you've got a family. That's reason enough not to be ambitious."

He poured himself a third glass. I thought I'd let it go at that.

"They are idiots!" he exploded after a prolonged silence. "They killed a lot of people."

Pavlov returned to Moscow after several more days of paperwork in Berdsk. He was careful never to mention Sverdlovsk again.

The story went public a few months later—in a way. In November 1979, a Russian magazine published by anti-Soviet émigrés in what was then West Germany reported that an explosion in a military facility in the southwest section of Sverdlovsk had released a cloud of deadly bacteria the previous April. It claimed that as many as a thousand people had died. Western news agencies picked up the story, quoting U.S. intelligence officials who claimed that the accident was clear evidence of Soviet violation of the 1972 Biological Weapons Convention.

Moscow denied the reports. On June 12, 1980, a statement published by the official Soviet news agency TASS declared that there had only been a "natural outbreak of anthrax among domestic animals" in the Sverdlovsk region.

"Cases of skin and intestinal forms of anthrax were reported in people, because dressing of animals was sometimes conducted without observing rules established by veterinary inspections," the statement said, adding that all of the patients had been treated successfully in local hospitals.

This was a lie, of course.

The German magazine and the U.S. intelligence sources were right that there had been an accident, but they got many of the facts wrong. Within a year, every senior Biopreparat official knew that something terrible had happened at Sverdlovsk. Nothing was said officially, but the news spread like wildfire. I learned the truth by talking to people who had been at the plant when the accident happened, and to army officers who had been in charge of the cleanup.

My pursuit of the facts was not a matter of indulging idle curiosity. We had to know what had happened if we were to protect ourselves from a similar disaster. As I rose higher in The System, I applied some of the lessons of Sverdlovsk to the plants under my control.

In fact, as neither Oleg Pavlov nor I could have known at the time, the Sverdlovsk incident would precipitate my speedy advancement. Not only did it help me get the "serious" work I'd been craving, but it set Biopreparat on a new course of development over the next decade.

On the last Friday of March 1979, a technician in the anthrax drying plant at Compound 19, the biological arms production facility in Sverdlovsk, scribbled a quick note for his supervisor before going home. "Filter clogged so I've removed it. Replacement necessary," the note said.

Compound 19 was the Fifteenth Directorate's busiest production plant. Three shifts operated around the clock, manufacturing a dry anthrax weapon for the Soviet arsenal. It was stressful and dangerous work. The fermented anthrax cultures had to be separated from their liquid base and dried before they could be ground into a fine powder for use in an aerosol form, and there were always spores floating in the air. Workers were given regular vaccinations, but the large filters clamped over the exhaust pipes were all that stood between the anthrax dust and the outside world.

After each shift, the big drying machines were shut down

briefly for maintenance checks. A clogged air filter was not an unusual occurrence, but it had to be replaced immediately.

Lieutenant Colonel Nikolai Chernyshov, supervisor of the afternoon shift that day, was in as much of a hurry to get home as his workers. Under the army's rules, he should have recorded the information about the defective filter in the logbook for the next shift, but perhaps the importance of the technician's note didn't register in his mind, or perhaps he was simply overtired.

When the night shift manager came on duty, he scanned the logbook. Finding nothing unusual, he gave the command to start the machines up again. A fine dust containing anthrax spores and chemical additives swept through the exhaust pipes into the night air.

Several hours passed before a worker noticed that the filter was missing. The shift supervisor shut the machines down at once and ordered a new filter installed. Several senior officers were informed, but no one alerted city officials or Ministry of Defense headquarters in Moscow.

In the next few days, all the workers on the night shift of a ceramic-making plant across the street from the facility fell ill. The plant had been directly in the path of the wind that night. Within a week, nearly all of them were dead.

By then hospitals were admitting dozens of patients from other areas of town who had worked in the plant's vicinity. Curiously, there were few women or children among the victims. Years later, some Western analysts wondered if the Soviets had developed a "gender weapon" capable of attacking only adult males. But women seldom worked night shifts in production plants, and few children would have been playing in the streets late on a Friday night.

Western scientists who have examined data from the accident believe that it occurred on Tuesday, April 3, or Wednesday, April 4, because the first cases did not surface until two or three days after that, which would fit the usual incubation period for anthrax. These arguments suggest to me how well Soviet officials were able to manipulate information and conceal the truth.

A colleague placed the accident on Friday, March 30, 1979. He was a Sverdlovsk scientist who recalled that he and other technicians learned of the first anthrax death—an auxiliary worker named Nikolayev—on the following Monday. That it happened on a Friday night helps explain why the workers were so anxious to get home and why so many people had passed by that evening, heading for a drink at a nearby bar. It is not unreasonable to assume that the KGB coverup included altering the dates on the medical reports of the first cases.

The last case was reported on May 19. The Soviet Union later claimed that 96 people were stricken with the disease and 66 died. The scientist who was working in the Sverdlovsk facility at the time told me the death toll was 105, but we will probably never know for sure. What is certain is that it was the worst single outbreak of inhalational anthrax on record this century.

There could have been no illusions in Moscow as to the cause of the outbreak. Chernyshov's lapse in judgment was reported as soon as the first deaths occurred. A delegation led by Colonel General Yefim Smirnov, commander of the Fifteenth Directorate, flew to Sverdlovsk a week after the incident. He was joined by Pyotr Burgasov, then deputy minister of health and a member of the Soviet Academy of Sciences. Burgasov brought with him a team of five doctors, but the government's concern for secrecy determined the handling of the medical crisis.

No one wanted to set off a panic or to alert outsiders. Sverdlovsk residents were informed that the deaths were caused by a truckload of contaminated meat sold on the black market. Printed fliers advised people to stay away from "unofficial" food vendors. More than one hundred stray dogs were rounded up and killed, on the grounds that they represented a danger to public health after having been seen scavenging near markets where the meat was sold. Meanwhile, military sentries were posted in the immediate neighborhood of the plant to keep intruders away, and KGB officers pretending to be doctors visited the homes of victims' families with falsified death certificates.

Whether residents suspected the truth or not, military and KGB control ensured that the city remained orderly. A Northwestern University physics professor named Donald E. Ellis, who was in Sverdlovsk at the time on an exchange program, reported that he noticed nothing unusual in the city. "I don't exclude the possibility that something may have occurred," he told *The New York Times* years later, "but I think either I or my wife would have sensed some effort to protect us from it. We . . . were not aware of any restrictions."

Residents had been living behind a thick veil of security for decades. Since World War II, Sverdlovsk, renamed after an early Bolshevik leader, had been the heart of the Soviet military-industrial complex, turning out tanks, nuclear rockets, and other armaments as well as biological weapons. In 1958, a major nuclear accident occurred at another site in the region, near the city of Chelyabinsk. The exact details of what happened are hazy, but reports from both Western and Communist sources indicate that a military reactor was damaged, resulting in the spread of radioactive dust over several thousand square kilometers. Twelve villages were evacuated.

The determination with which Soviet officials set about concealing the Sverdlovsk leak from their own people as well as the world was, under the circumstances, not surprising. The truth would have severely embarrassed the nation's leaders, many of whom were not even aware that biological arms production was under way, and caused an international crisis. It wasn't at first clear that the coverup would succeed. Army commanders worried that they might not be able to contain the disaster.

"We couldn't understand why people continued to die," a general who was there told me much later. "We assumed that this was a quick, one-time exposure and that our mopping-up would be completed in a few days, but there were deaths for a month and a half after the release."

The coverup was responsible for turning what began as a medical emergency into a small epidemic.

The local Communist Party boss, who was apparently told that there had been a leak of hazardous material from the plant, ordered city workers to scrub and trim trees, spray roads, and hose down roofs. This spread the spores further through "secondary aerosols"—spores that had settled after the initial release and were stirred up again by the cleanup blitz. Anthrax dust drifted through the city, and new victims arrived at the hospitals with black ulcerous swellings on their skin.

The cutaneous form of anthrax, contracted when spores enter the body through a cut or abrasion on the surface of the skin, occurs naturally in rural areas around the world, especially those with large herds of domestic cattle, sheep, and goats. It is the most common form of anthrax and is rarely lethal when treated with antibiotics such as penicillin. Russians refer to it as the "Siberian ulcer," as it manifests itself through the formation of small and localized lesions on the surface of the skin. An outbreak of cutaneous anthrax in the region was credible, but it wouldn't explain why so many factory workers, who could have had no contact with animals, were suddenly sick, or why so many died.

Anthrax spores can survive for years—even decades—in a dormant state. Animals will become infected while foraging for food. The spores germinate in a matter of hours and multiply in their hosts, returning to spore form when they die or on contact with oxygen. Men and women who work with infected animals—butchers, tanners, farmers, and workers in textile mills—become infected in turn through abrasions, by inhaling spores or drinking contaminated water, or, in rare cases, by eating contaminated meat.

Soviet officials fell back on the claim that the disease was caused by contaminated meat. Doctors displayed photos that suggested victims had contracted intestinal anthrax, by far the rarest form of the disease (it accounts for fewer than 1 percent of all cases). But officials could not hide the presence of pulmonary or inhalational anthrax, the most lethal of all.

Between ten thousand and twenty thousand spores, a microscopic quantity, are sufficient to infect someone with anthrax. The same anthrax bacterium will behave differently depending on how it enters the system. It is far more threatening if it is inhaled or ingested than if it enters through the skin. Inhalational anthrax was first identified in the early nineteenth century when workers in a textile mill were exposed to spores released into the air by the new industrial processes developed to make wool. It is often called wool sorters' disease.

As soon as an anthrax spore enters the body it germinates and begins to multiply. A few days will pass before the anthrax bacteria produce toxins which, in the simplest terms, bind to the protective membranes of target cells and cripple the ability of white blood cells to fight off disease. It is the toxin, and not the bacterium itself, that ravages the body and is responsible for death. If an anthrax victim is treated with high doses of penicillin injected into the bloodstream at short intervals for a week to ten days before the first toxins are released, chances of survival are almost 100 percent. But antibiotics can do little to fight the anthrax toxin. Combinations of penicillin and streptomycin have been used at this stage, but the prognosis is grim.

The headlong trajectory of pulmonary anthrax can be blocked if penicillin is administered before the first symptoms appear. I was told that thousands of Sverdlovsk residents were given antibiotics and vaccinated immediately after the first cases were reported, but it was too late to save the victims who had already begun to suffer from the fever, shortness of breath, and distinctive dark swellings along their chest and neck that mark the onset of pulmonary anthrax.

Sverdlovsk's anthrax was the most powerful of the dozens of strains investigated over the years by army scientists for their weapons potential. It was called Anthrax 836 and had been isolated, ironically, after another accident.

In 1953, a leak from the Kirov bacteriological facility spread anthrax into the city's sewer system. Vladimir Sizov, the army biologist who discovered the strain, came to work for Biopreparat years later and told me the story.

According to Sizov, an unknown quantity of liquid anthrax was accidentally released by a defective reactor at the Kirov plant. Army workers disinfected the sewer system immediately but soon found evidence of anthrax among the rodent population. Disinfections were ordered regularly after that, yet the disease continued to lurk underground for years. In 1956, Sizov found that one of the rodents captured in the Kirov sewers had developed a new strain, more virulent than the original. The army immediately ordered him to cultivate the new strain. It was eventually used as the basis for the weapon we planned to install in the SS-18s targeted on Western cities.

If it is impossible today to reconstruct exactly what happened during those frantic weeks in April and May of 1979, this is in part because the KGB did its work so well. I was told by army personnel involved in the cleanup that the corpses of the victims were bathed in chemical disinfectants and that much of the documentary evidence, including hospital records and pathologists' reports, was destroyed. To add verisimilitude to the cover story, several black-market vendors in Sverdlovsk were imprisoned on charges of selling contaminated meat.

I have often wondered whether the Party boss who ordered the rapid cleanup understood the fatal consequences of his actions. He should certainly be asked. The Communist Party chairman of Sverdlovsk at the time of the accident was Boris Yeltsin, the first leader of post-Soviet Russia.

Smirnov, the Fifteenth Directorate commander, met daily throughout the crisis with Yeltsin, a hard-nosed former construction manager who fought his way up the Party ladder to become head of the region, a position equivalent to the governor of an American state. Yeltsin enjoyed a reputation as a blunt politician who enjoyed putting the area's petty military-industrial tyrants in their places. He was as loyal as any other apparatchik to the Communist system and was keenly aware that he was expected to keep the regime's secrets. As Party chief of Sverdlovsk, he had carried out a Kremlin order to bulldoze the house where Czar Nicholas II and his family had been murdered in 1918.

According to a high-ranking military official who was in Sverdlovsk at the time, Yeltsin was so enraged by the lack of cooperation he received that he stormed over to Compound 19 and demanded entry. He was refused, on the orders of Defense Minister Dmitry Ustinov, who took over when Marshal Grechko died in 1976. Ustinov arrived at the site two weeks after the accident. As a Politburo member, he far outranked a provincial party boss.

While Yeltsin has since apologized for his role in bulldozing the house where the czar was murdered, he has said almost nothing about the anthrax accident at Sverdlovsk. In his autobiography *Against the Grain*, published in 1990, he briefly referred to the "tragic" outbreak and tucked away in a footnote the assertion that the epidemic was caused by a "leak from a secret factory." A full accounting is long overdue.

In the years since the accident, Sverdlovsk has been called by analysts and many Russians themselves a "biological Chernobyl." They are right. The casualty figures do not compare with those following the 1986 explosion at the Ukrainian nuclear plant, but just as the Chernobyl disaster alerted the world to our questionable management of nuclear power, Sverdlovsk was a grim warning of the dangers of our secret science.

In February 1981, two years after the accident, I received an anxious call from the director of Omutninsk, Vladimir Valov, at my office in Building 107. I was then chief of the main technological department. The message was that some "very important generals" would arrive at the compound later that day.

"Tell everyone in your staff to go home early—except for the technicians in the inner zones," he said. "You'll have to stay around to escort them."

At 5:30 P.M. a jeep pulled up to the front door and two officers emerged. The first was General Vladimir Lebedinsky, who had replaced Smirnov as Fifteenth Directorate commander. The second officer was taller and heavyset. He had a distinctive air of authority, suggesting he was senior to Lebedinsky. Later, I was told he was the head of the military department of the Communist Party Central Committee, the real source of power in the military establishment. His name was Shakhov. Both men wore civilian clothing.

Lebedinsky was surprisingly polite for a senior officer. He apologized for inconveniencing me and asked if I could show them Building 107. Everyone was curious about Omutninsk, and I was getting used to conducting tours for top officials. I brought them proudly to a spot where they could look through sealed windows into Zone Two, the first biosafety enclosure.

They peered at the storage vaults and rows of seed and industrial reactors. One or two hooded technicians were doing cleanup work.

"That's a hell of a lot of glass for such a small amount of germs," Lebedinsky joked. "We don't need to go through all of that in our installations."

His partner gave him a cold look.

"Maybe if you had, Comrade General, you would have avoided what happened at Sverdlovsk," he said quietly.

Lebedinsky turned pale and said nothing. I had never seen a powerful general cut down like that. Both men seemed to have forgotten I was there.

After a few moments Lebedinsky turned on his heels and walked past us down the corridor. I was about to follow, but Shakhov put his hand on my arm and shook his head. He followed the general out a few moments later, ending our tour.

The biggest challenge facing the biological warfare establishment after Sverdlovsk was what to do with the plant itself. It couldn't continue anthrax production now that the eyes of the West were fixed on its activities. The city was closed to foreigners, but we could be certain that Western surveillance efforts would increase.

Three facilities in the country were designated as centers for anthrax production in case of war: Sverdlovsk, Penza, and Kurgan. Sverdlovsk had been the only active production facility; the others were on standby, keeping strains of anthrax in their vaults for the day when an order from Moscow would activate their production lines. The army was desperate to get the industrial anthrax production lines at Sverdlovsk running again. It lobbied hard to revoke the temporary suspension of activities at the plant ordered by Party bosses after the accident.

Pressure to produce more biological weapons was increasing by the month, but hardly anyone at senior levels of the government understood what they were. The average military commander regarded biological armaments as another type of weapon, slightly more useful than dynamite, perhaps, but not particularly more dangerous. Party bureaucrats recognized how lethal such weapons could be, but they didn't understand the unique hazards associated with making them.

Biopreparat took advantage of this confusion to press its case. Our tularemia weapon had shown that we could be as successful as the army in developing new weapons. The fact that we were outwardly a civilian organization made it more likely that our work could be concealed from the West. To the army's astonishment, it found itself outflanked by the tiny agency it had once regarded with contempt.

In 1981, Brezhnev signed a secret decree ordering the relocation of all biological weapons-making equipment and materials from Sverdlovsk to Stepnogorsk, a small biological research facility operated by Biopreparat in the remote deserts of northern Kazakhstan.

This decision would affect me directly. I was developing a reputation for getting results. Uncertainties about the direction of my life, and the morality of what I was doing, had long since receded.

Everyone soon learned of the plans to upgrade Stepnogorsk for anthrax production—it was the subject of office gossip for months. The most ambitious amongst us were eager to be part of a project that was bound to receive unlimited support and money. I put in a bid to become manager, the warning of Oleg Pavlov having long since been forgotten.

I was still only a major, a rank that didn't qualify me for a senior management position, but I was filled with more confidence than I probably deserved. My success with tularemia had given me an edge over other candidates, and I thought I could handle the job.

I called Kalinin and asked to speak with him about taking over as director of the plant. I think he enjoyed the brashness of my approach. There was just one difficulty: Stepnogorsk already had a director, a colonel appointed earlier that year. Kalinin told me to go on vacation while he considered a strategy.

A few weeks later I was on the way to Stepnogorsk with my family. I had been appointed deputy director of the upgraded anthrax facility. After we settled in, I went out for dinner with the director and some of the Stepnogorsk managers. Toward the end of the evening, the vodka was running freely.

The director, a colonel named Davydkin, pulled me to his side and gave me a playful punch.

"Kanatjan," he said, "it's really nice to have you here—but I want you to tell me the truth. You're here to take my job, aren't you?"

I laughed. "Of course not! Where did you ever get that idea?"

In less than a month, Davydkin was transferred and I was appointed director of the Kazakhstan Scientific and Production Base in Stepnogorsk.

Back in Sverdlovsk, anthrax production at Compound 19 was

My grandfather Abdrahman Aitiev at his desk in the 1920s. One of the leaders of the Communist revolution in Kazakhstan, he was the first People's Commissar of Internal Affairs, responsible for police and security services in the 1920s and 1930s. A Kazak national hero, he was imprisoned in 1936 and died that year in a prison hospital in mysterious circumstances. Several streets in Kazakhstan are named after him.

My parents, Rosa and Byzak Alibekov, in 1950, when my father was a junior police lieutenant and my mother was pregnant with me.

Taking the army oath to enter the military faculty of the Tomsk Medical Institute in 1973.

A break from military training with fellow medical cadets in 1974. I am second from the left. To my right is Talgat Nurmagambetov, recent chief of medical services for the Kazak army.

Military training: laying a simulated mine at Tomsk in 1974.

Delivering a report on the tularemia outbreak at the Battle of Stalingrad at a scientific conference for military cadets in early 1975.

With my platoon at graduation: June 1975. I am bottom row, second from the right. Top row on the far left is Yevgeny Staroverov. He was assigned with me to Omutninsk, where he still worked in the late 1980s. In the second row, second from the left is Yevgeny Stavsky, who joined the Fifteenth Directorate after graduation and went to Vector in the 1980s as a department chief to develop a smallpox weapon.

With my first child, Mira, at Berdsk in 1979, when I was working with brucellosis and was accused by the KGB of illegally developing biological weapons.

An official army photo, taken in 1982, after I was promoted to deputy director of Omutninsk. I am wearing a medal for "wartime services" awarded for the successful development of a tularemia biological weapon.

May Day at Stepnogorsk, 1985. Anthrax weapons developers and their children.

Part of the Stepnogorsk compound: a view from Building 221. The building in the foreground housed aerosol explosive chambers. Directly behind it is an anthrax drying facility. The bunkers in the foreground were for filling and assembling biological munitions.

With my family in August 1987. From the right: Lena, Alan, Mira, Timur. This photo was taken after the successful testing of my new anthrax weapon, a month before we moved to Moscow.

General Kalinin presides: an annual conference of bioweaponeers—institute directors, scientists, and chief engineers—near Moscow in late 1988. Right to left: General Nikolai Urakov, director of Obolensk; me; Oleg Ignatiev, chief of Biological Weapons Directorate of the Military Industrial Commission (VPK); unidentified official; and Kalinin.

General Kalinin's audience.

First doubts: with Colonel Professor Tarumov, one of the main developers of tularemia biological weapons, at a scientific conference in Moscow in 1990.

Military coup: Biomash employees at the barricades in front of the Russian White House in August 1991.

Pine Bluff, Arkansas: inspecting American facilities. Sitting, from left to right: General Urakov; GRU colonel Dzuby; Lev Sandakchiev, director of Vector; Colonel Shcherbakov, chief of the scientific directorate at Biopreparat; Lisa Bronson, head of the American delegation; me. Third from the right, standing, is Colonel Vasiliev, deputy commander of the Fifteenth Directorate. Standing in front of the pole is Grigory Berdennikov, head of the Russian team. He was appointed deputy minister of foreign affairs in 1992.

Giant autoclave in Building 221 at Stepnogorsk, used to sterilize nutrient media and for the deactivation of anthrax cultures.

Dismantling Stepnogorsk: Pilot plant installation for filling and sealing biological bomblets.

Fort Detrick, Maryland, 1998: with three successive commanders of USAMRIID—Charlie Bailey, Dr. Dave Franz, and Dr. Jerry Parker.

Fort McClellan, Alabama: a briefing at the U.S. army biological weapons reconnaissance installation. The truck behind us is equipped to detect and identify biological weapons.

AP Photo/Gino Domenico

New York, Sunday November 9, 1997: Emergency workers evacuate a "victim" of a biological weapons attack in a training exercise a few blocks from City Hall.

Visiting Charlie Bailey at home in Alabama, 1998.

officially stopped. The military facility would continue to serve as a research base and a storage site for biological weapons. In 1983, the first of sixty-five army technicians and scientists from the now-discredited anthrax plant in the Urals began to arrive in Kazakhstan. One of them was Nikolai Chernyshov.

Chernyshov walked into my office in 1984, accompanied by the chief of Stepnogorsk's Biosafety Division, a lieutenant colonel named Gennady Lepyoshkin. The two young men made a startling contrast.

Lepyoshkin was sharp-tongued and gregarious, a man whose energy made him seem larger than any room he occupied. Chernyshov was a little older, in his late thirties, his brown hair already flecked with gray. I knew nothing about his background except that he was regarded as an expert in anthrax drying methods. He refused to look me in the eyes.

I hated to act as a "boss," especially in the company of men my age, and we soon launched into a free-flowing discussion about people we knew in common at Biopreparat and the Fifteenth Directorate. Chernyshov hardly participated. I noticed that his hands trembled as he held his teacup.

Lepyoshkin noticed my glances and began to grin.

"Kolya!" he said, turning to Chernyshov. "Why don't you tell our commander Kanatjan what you have done?"

"Go on, tell me," I said with a smile, enjoying our camaraderie. "I won't punish you."

I thought Chernyshov might have committed some embarrassing blunder in the lab. He was an experienced scientist; I couldn't imagine it was anything serious.

Chernyshov turned beet red. He kept sipping his tea and refused to talk.

Lepyoshkin was enjoying himself too much to hold back.

"Have you heard about the Sverdlovsk accident?" he asked me.

By then, of course, I had.

"Do you know who was responsible?"

"Who?"

"You're sitting across the table from him."

I stared at Chernyshov in disbelief. His face was riveted on an invisible spot in front of him, and his hands began to shake so violently that he had to put his teacup down. He looked as if he was about to burst into tears.

Lepyoshkin began to describe what had happened that March afternoon in Sverdlovsk. Chernyshov didn't try to deny a thing. He refused to say a word.

His friend kept smiling. "So, now you know: this is the guy who killed all those people."

Chernyshov finally got up and walked out.

I thought Lepyoshkin had been unnecessarily harsh, but I also felt a stirring of anger. Chernyshov would carry the guilt for his moment of thoughtlessness for the rest of his life. But he had never been punished, and no one in any position of authority had bothered to inform me of his responsibility for the accident before he was transferred to my facility.

For the good of our biological warfare program, Chernyshov's mistake had to be kept quiet. A thorough investigation of what had happened in Compound 19 would raise too many awkward questions even inside our own government about our activities. This was further proof that secrecy was valued above all else in our system—even if it endangered our own safety. In the West, an accident of such magnitude would have been investigated ad nauseam and its lessons distributed, however quietly, to those working in similar areas. Our coverup virtually guaranteed further disasters.

A few months later I ran into another veteran of Sverdlovsk, Lieutenant Colonel Boris Kozhevnikov. In the year following the accident, he told me, a work crew was ordered to take a boxload of 250-liter containers filled with dried anthrax to storage bunkers inside Compound 19. Kozhevnikov had been assigned to escort the workers as they rolled the containers on carts toward the bunker a few hundred feet away. One cart hit a bump, and a container fell open.

I was aghast. "What did you do?" I asked him.

"I just closed it." He shrugged.

Hastily, he added that he had ordered disinfectant poured everywhere. No one had fallen sick. And, of course, his superiors were not informed.

Nine years after the Sverdlovsk accident, a group of Soviet medical experts arrived in the United States to reveal the "truth" about what happened in 1979. Invited by Dr. Matthew Meselson, a noted Harvard professor, they toured Washington, Baltimore, and Cambridge with a stack of reports and photographs purporting to show that all the victims had contracted either intestinal or cutaneous anthrax. Pyotr Burgasov, who led the original Ministry of Health team into Sverdlovsk, headed the delegation.

Burgasov had by then retired as deputy health minister to become a government adviser. With a rueful smile, he acknowledged that a public explanation was long overdue. He blamed the delay on the Soviet government's reluctance to reveal embarrassing deficiencies in its public health system. The West's fascination with *perestroika* and *glasnost* helped persuade most of his listeners.

"Sverdlovsk's 'mystery epidemic' of 1979 lost much of its mystery this month," declared the respected U.S. journal *Science* in an April 1988 account of the Soviet doctors' trip. "For eight years, U.S. officials have voiced suspicions about an unprecedented outbreak of anthrax that occurred in April 1979 among the people of Sverdlovsk; [but the Soviets claimed] people had become sick . . . from eating bad meat they bought from 'private' butchers.

"Three Soviet officials came to visit the National Academy of Sciences in Washington, D.C., on 11 April . . . [they] gave the same explanation as in 1980, but provided many more details, convincing some long-time doubters that the account was true."

A few months before Burgasov and the other officials left for the United States, a copy of the paper that he was to present in America landed on my desk in Moscow. I was asked, as Biopreparat's scientific chief, to rubber-stamp his conclusions.

At the time, it didn't matter to me whether Americans were told the truth or not, but I thought Burgasov's account would never pass muster with any self-respecting epidemiologist. How could anyone believe that people would go on eating "contaminated" meat for weeks after the first victims fell sick? The story might explain a few deaths, but not an epidemic. And how could they explain that the majority of the victims were adult males? Didn't women and children eat meat?

The man who had asked for my comments was General Lebe-dinsky. When I returned to his office at army headquarters, I handed the paper back to him.

"Can you tell me, General," I said, "what the real cause of the Sverdlovsk accident was?"

"Contaminated meat, of course," he said at once.

I reminded him of the afternoon in Omutninsk, years earlier, where he had been reprimanded by the man from the Central Committee.

He looked surprised.

"You remember that?" he said. He flashed one of his paternal smiles.

"Listen," he said. "If you think you know what caused this thing that's your business, but never ask me what happened. Each time you ask, my answer will always be 'contaminated meat.' "

I refused to sign off on the paper, believing it would make us look foolish abroad. Burgasov was furious.

"Tell that young man to write his own paper," he fumed at Kalinin, who impolitically passed the remark on to me.

Burgasov took his version of events to America, and I was astonished to hear that his visit was a success.

The truth about Sverdlovsk, or at least some of it, finally emerged in Russia during an interview granted by Boris Yeltsin to a *Komsomolskaya Pravda* reporter, published on May 27, 1993.

"Our military developments were the reason [for the accident]," Yeltsin said cryptically, adding that he had asked then–KGB chairman Yuri Andropov and Defense Minister Ustinov to close down the bacteriological facility as soon as he heard about the anthrax release.

When the reporter asked why he had been silent for so long, Yeltsin answered, "Nobody asked me."

Now, for some reason, the coverup has resumed. In 1998, Russian newspapers published articles quoting officials on the "real" cause of the anthrax outbreak two decades earlier.

They said it was contaminated meat.

8

PROGRESS

Stepnogorsk, 1983–87

Within a few weeks of my assignment to Stepnogorsk I was ordered to Moscow for briefings. It was a sobering experience. At the KGB's First Department, on the top floor of the Samokatnaya Street headquarters, I was shown a secret decree issued by Brezhnev the previous year, in 1982. I had never been allowed to see such top-secret material before.

An intelligence officer pulled the decree from a red folder tied with a string, placed it gravely on a desk, and stood behind me while I read. He would only let me see the sections that corresponded to my duties. I already knew the gist of the order: we were to transform our sleepy facility in northern Kazakhstan into a munitions plant that would eventually replace Sverdlovsk.

Anthrax 836, first discovered in Kirov in 1953, was our best candidate to become what we called a "battle strain"—one that was reproducible in large quantities, of high virulence, and transportable. Once I'd worked out the technique for its cultivation, concentration, and preparation, I was to develop the infrastructure to reproduce it on a massive scale—a goal that had eluded our military scientists for years. This meant assembling batteries of fermenters, drying and milling machines, and centrifuges, as well as the equipment required for preparing and filling hundreds of bombs.

My job at Stepnogorsk was, in effect, to create the world's most efficient assembly line for the mass production of weaponized anthrax.

There were many in Moscow's close-knit biological warfare establishment who believed it couldn't be done and who hoped that Biopreparat and its assertive commander would stumble in the effort. Our success with tularemia the previous year had turned Kalinin into an influential figure. He was ruffling egos throughout the army command.

At one of my Moscow meetings, an elderly general named Tarasenko, then deputy commander of the Fifteenth Directorate, pulled me aside.

"Congratulations on your new job, Major Alibekov," he said, patting me on the shoulder. "It's about time we gave our young people more responsibility."

I smiled, pleased by his attention. Tarasenko was a veteran military scientist, one of the most respected figures in Soviet biological and chemical weapons research.

"But you should watch out for yourself," he continued. "The mountain of metal they want you to build down there will never do what it's supposed to do. Believe me, I've had thirty years of experience with these things, and I know what works. This won't."

I was too stunned to reply. He nudged my shoulder again.

"I'm sure you'll be given everything you need," he said, "but Biopreparat is trying to create a monument, and in the end you'll be the one who will have to dismantle it when it falls."

The Kazakhstan Scientific and Production Base was established in 1982, tacked on to a state enterprise called the Progress Scientific and Production Association, which manufactured pesticides and fertilizer. The new facility occupied more than half the buildings in the compound, but the several thousand pesticide workers employed there could not be told of its new function. It had been chosen as one of six biowarfare facilities in the country designed to be mobilized as special production units in the event of war.

Officially, I was deputy director of the Progress Association, but my secret job title gave me more authority than the director: I was "war commander" of the entire installation. This was a daunting prospect for someone whose military knowledge was based on two years of basic training. I was expected to take control of the factory during what the army called "special periods" of rising tensions between the superpowers. Upon receipt of a coded message from Moscow, I was to transform Progress into a munitions plant.

Strains of virulent bacteria would be pulled from our vaults and seeded in our reactors and fermenters. Anthrax was our main agent at Stepnogorsk, but we also worked with glanders and were prepared to weaponize tularemia and plague. The pathogenic weapons that emerged would be poured into bomblets and spray tanks and loaded into trucks for shipment to a railroad station or airfield, from which point they would be transported to military sites around Russia for placement on bombers or ballistic missiles.

I was to maintain production until I received an order from Moscow to stop, or until our plant was destroyed.

It may be hard for anyone to imagine today the seriousness with which we prepared for war, but along with most of my colleagues, I believed that a superpower conflict was inevitable.

In the early 1980s, relations between the Soviet Union and the West had plummeted to their lowest point in decades. The election of President Ronald Reagan had led to the biggest American arms buildup our generation had seen. Our soldiers were dying in Afghanistan at the hands of U.S.–backed guerrillas, and Washington was about to deploy a new generation of cruise missiles in Western Europe, capable of reaching Soviet soil in minutes. Intelligence reports claimed that Americans envisioned the death of at least sixty million Soviet citizens in the case of a nuclear war.

We didn't need hawkish intelligence briefings to persuade us of the danger. Our newspapers chafed over Reagan's description of our country as an evil empire, and the angry rhetoric of our leaders undermined the sense of security most of us had grown up with during the détente of the 1970s. Although we joked amongst ourselves about the senile old men in the Kremlin, it was easy to believe that the West would seize upon our moment of weakness to destroy us. It was even conceivable that our army strategists would call for a preemptive strike, perhaps with biological weapons.

The Progress Scientific and Production Association was the eeriest place I had ever worked. Encamped on a windy plain ten miles from the uranium-mining town of Stepnogorsk, the enterprise was ringed with high gray walls and an electric-wire fence. The surrounding land was stripped of all vegetation, partly as a safeguard against the accidental release of pathogens (a lesson from Sverdlovsk, where hosing down the contaminated bushes had created a new source of infection), and partly to preserve a clear line of sight against intruders. Motion sensors were embedded everywhere.

Inside the compound, dozens of white and gray buildings were arranged on a grid of narrow streets. It was a miniature city, with a skyline of oddly shaped towers and buildings, some more than five stories high. There were separate entrances for civilian and military employees. Armed guards were stationed at both.

Security inside the compound was even more oppressive. Following the Sverdlovsk accident, intelligence organs had increased their influence throughout the Soviet biological weapons establishment. No explicit connection was ever drawn between what had happened in the Urals and the tightening of security regulations, but it was made clear to us that no one wanted another uncomfortable bout of international attention.

As a lab scientist, my worst fear had been that a careless act might put my life and that of my immediate coworkers at risk. As director, I had become responsible for the health and safety of tens of thousands of people in the nearby community. It was my job to make sure that our secrets stayed, both literally and figuratively, behind the walls of our compound.

At night, Lena told me I was grinding my teeth so hard it kept her awake. And I was talking in my sleep.

"You keep mumbling about requirements for this, requirements for that," she smiled. "You ought to forget about things when you come home."

Even if I had wanted to forget, one person at Stepnogorsk was determined to keep my level of anxiety as high as possible.

One morning in October 1983, a few weeks after I arrived, KGB lieutenant colonel Anatoly Bulgak, commander of the facility's counterintelligence unit, poked his head around the door of my office.

"Mind if I come in?" he said.

He was inside before I could reply.

He plumped down in a chair close to my desk and casually stretched his legs out on the floor as if he had been doing this every morning of his life.

"Since you and I have to work together," he said, "I think we should be friends."

"What do you need?" I said.

"It's not what I need. It's really what you need."

He paused. I said nothing.

"You and I know a lot of new people will have to be brought here in the near future," he went on, clearly disappointed by my lack of response. "This place will get crowded, and it will be easy to make mistakes."

"Mistakes?"

He sat up in the chair and put a hand on each knee, to emphasize the seriousness of what he was about to say.

"Mistakes of the security kind. We can't afford that."

I bristled at his use of "we."

"I don't see any reason to worry," I said. "I'm sure I'll be able to call on you if there's a problem."

"You don't understand," Bulgak went on impassively. "It would be natural, and sensible, to go over the personnel lists ahead of time. That way, there won't be any problems with Moscow."

"I don't think you are qualified to choose scientists—are you?" I said.

I tried to make this sound like an innocent remark, but he recognized it immediately as the insult it was meant to be.

Bulgak possessed a country policeman's dim awe of science. He resented anyone who made him conscious of his ignorance. When I came to know him better, I understood that he was terrified of

what went on in our labs. Whenever he suspected a worker of pil-
fering, or protested one of our security procedures, I would invite
him to don a biological protective suit and come with me into
Zone Two so as to investigate for himself, but he would back off,
hastily insisting he would join me "some other time" when he
wasn't busy.

A knock on the door broke the silence. It was one of my lab
chiefs, with a problem demanding immediate attention. Bulgak
rose to his feet and smirked.

"You can't do this without my help," he said. "Try it, you'll
see."

He sidled past my visitor without acknowledging his presence
and walked out. I caught a faint look of distaste in the lab chief's
eyes as we watched him go.

Bulgak was easy to dislike. He was a bland-featured man in his
mid-thirties with shrewd eyes and an unpleasant demeanor. Every
element of his personality appeared calculated to impress others.
The clothes he favored—padded ash gray suits and dark shoes—
seemed to have been chosen from a catalog for secret policemen.
He had been transferred to Stepnogorsk from a rural KGB office in
southern Kazakhstan six months before my arrival.

But Bulgak was smaller than the sum of his parts. While plant
workers feared him in person, they mocked him behind his back. I
didn't need to know much about him to understand the power he
commanded in our institution. The KGB operated a counterintelligence unit in every biological weapons research lab in the Soviet
Union. Its chief automatically served as a deputy director of the facility, but he reported through his KGB superiors to Lubyanka, the
massive building in central Moscow that had served as the nation's
secret police headquarters since the early years of Soviet power.

Every director had to accept this alternative chain of command
without complaint. The KGB devoted as much energy to watching
Biopreparat's senior managers as it did to lower-ranking employees. Avoiding intelligence scrutiny was impossible: although fewer
than ten or fifteen KGB employees were assigned to each facility,
the units relied on informers to keep us in line.

Savva Yermoshin used to boast that "one out of every ten Soviet citizens" unofficially reported to the security organs, insinuat-

ing that the same ratio held true in our agency. I never tried to dispute him. He had probably already discovered from my personnel file that I nearly became a KGB informer myself. My bitter memory of that experience no doubt influenced my handling of Anatoly Bulgak.

It was in 1978, five years before I arrived at Stepnogorsk, when I was a junior scientist at Berdsk. I had just completed my first major assignment, the development of a lab technique for weaponizing brucellosis. The task had been authorized by Lev Klyucherov, who was then a colonel and Biopreparat's scientific chief, and in a burst of youthful exuberance I fired off a triumphant report to Moscow. I was sure Klyucherov would want to know the results immediately.

I received no answer. This should have been a warning.

A few days later, the commander of the KGB's counterintelligence unit at Berdsk, Colonel Filipenko, walked into my office holding a copy of my report.

"What does this mean?" he asked.

Flattered that even the KGB was interested in my work, I launched into a description of the steps I had taken to manufacture the weapon prototype. I was in the midst of a lengthy recital of the composition of the nutrient medium when Filipenko cut me off.

"I don't think you understand me," he said. "What I'm asking you is who told you to do this?"

Surprised, I said it was an assignment from Colonel Klyucherov.

"That can't be true," he snapped. "I just spoke to him in Moscow, and he knows nothing about it."

"But he gave . . ." I stopped in mid-sentence.

With a sinking feeling, I suddenly realized I had overlooked a regulation requiring us to inform the KGB detachment at our labs of all "special projects." This was a legacy of the prewar era, when the secret police, under Lavrenty Beria, were the coordinators of all biological warfare activity. Before the Sverdlovsk accident, most people ignored such minor security precautions, but the regulation was there to be enforced, and I had landed my superiors in a mess.

My supervisor at Berdsk was not sympathetic when I called him.

"You know you weren't supposed to create anything, just to analyze whether it could be done," he said coolly. "You went beyond your orders."

It was futile to argue. I realized that the KGB would seize on my indiscretion to charge that a Biopreparat scientist was developing weapons on his own. A good manager might have made excuses for my inexperience, but Klyucherov and the supervisor were more interested in protecting themselves and shielding Biopreparat.

The next day I was summoned to Berdsk KGB headquarters. A second notice was sent to Vladimir Rumyantsev, the friend who had worked with me on the project.

We walked to the KGB's two-story building near the center of town, too frightened to speak. An officer in civilian clothes escorted us inside and waved me into the commander's office. Rumyantsev was told to wait his turn.

The commander, Kuznetsov, was reading my report. Piles of paper were strewn across his desk. I looked around for somewhere to sit, but there was no chair.

Kuznetsov didn't bother to glance up when I walked in. He read my paper with an impassive absorption that reminded me of one of my professors at Tomsk and shook his head every few minutes in theatrical dismay. Finally, he pushed his chair back, stood up, and strode over to me, placing his face within a few inches of my own.

"Why did you do it?" he shouted.

"I received an order," I responded weakly.

"So, you're a fascist?"

"What?"

"Only a fascist would answer that he killed people because of an order."

"But I didn't kill anybody," I protested. "I just did the work I was asked to do."

"That doesn't matter. You are obviously the kind of person who would kill on demand. You have no brain of your own!"

His voice increased in decibels with each sentence. I was petrified. I almost began to believe I had killed someone.

The tirade seemed to go on for hours. Kuznetsov continued to accuse me of being a fascist, and I continued to deny it. I didn't know what else to do, or exactly what he wanted. Would he stop shouting if I confessed to my fascist tendencies? A confession seemed pointless, especially if I was going to be fired anyway.

The image of my father and his battle decorations passed through my mind. Would he believe his son was a fascist?

"Look," I said at last, my voice rising in frustration. "If you think I'm a fascist, why don't you put me in jail!"

Kuznetsov stopped shouting, looked deep into my eyes, and went back to his desk. The sudden quiet was chilling.

"Well, Lieutenant," he smiled, "we don't need to do that. People make mistakes. I can forgive you, perhaps—but I need your help."

"How?"

"I'll tell you," he said, spreading his large hands on the desk. "One of the things in your favor is that we know you've just joined the Party, correct?"

I nodded. Communist Party membership wasn't essential to employment in our labs, but it was one of the criteria for getting ahead. I had joined because I knew it would look good on my record.

"You're bright and you're a scientist," Kuznetsov continued, now radiating benevolence. "But a lot of other scientists haven't joined the Party. That means we really don't know what kind of people they are or what they're thinking. Maybe they have doubts; maybe some of them express opinions against our Motherland."

Kuznetsov looked at me expectantly, but I had no idea what he was getting at.

"Well," he came to the point. "You can help us figure out what's going on."

Then I understood. "You want me to be an informer?" I said.

"No, no," he answered quickly, as if the idea repelled him. "Just a kind of assistant."

All at once, my confidence returned. They weren't going to fire me, after all. I felt ashamed of having been so frightened. I replied, lightly, "Without pay?"

It took a moment for my question to sink in. Then Kuznetsov exploded again.

"If you think this is a joke, you'll be sorry soon enough," he said, and dismissed me.

I found Rumyantsev, pale and nervous, pacing the hallway outside the office. I didn't know what he had heard and was about to whisper an encouraging word when Kuznetsov appeared behind me.

"Don't wait around for your friend," he said. "Go home."

Rumyantsev came to my apartment later that night, carrying two bottles of vodka. Silently, we finished one bottle and then started on the other.

At last, he spoke.

"Kanatjan," he said, "I know you told them no."

"That's right," I answered, now fully recovered from Kuznetsov's interrogation and proud of having stood up to him.

He took another drink and pursed his lips. "It's the same thing I told them."

"It doesn't matter," I said. "Don't worry about it." We went on to other subjects.

Over the next several months he would from time to time pull me aside at parties and make oblique references to our interview with the KGB.

"You're such a good guy," he said once, clapping me on the back. "I'm a bad guy."

I didn't want to believe that Kuznetsov had been able to bully my friend into serving as an informer, but I also didn't want to know if it was true. We drifted apart after Berdsk. I helped him secure a senior post in Biopreparat when I became deputy director, but he was fired when his superiors accused him of arrogance. Many years later, after I arrived in the United States, I was hurt to learn that he had told a mutual friend I was a spy.

By the time we finished our construction program at Stepnogorsk, the facility looked like the mountain of metal Tarasenko had warned me about.

New buildings had risen from the desert floor, dramatically changing the skyline. One of them, Building 600, was the largest indoor testing facility constructed up to that time in the Soviet Union. It was more than fifty feet high, with two giant stainless steel testing chambers hidden inside. The first chamber, designed to withstand the force of a powerful explosion, would be used to analyze the decay rate and dissemination capacities of aerosol mixtures contained in our germ bombs. The second was for testing animals. We also constructed a network of underground bunkers to store our materials and an elaborate system of ventilation and waste pipes.

Bioweapons are not rocket launchers. They cannot be loaded and fired. The most virulent culture in a test tube is useless as an offensive weapon until it has been put through a process that gives it stability and predictability. The manufacturing technique is, in a sense, the real weapon, and it is harder to develop than individual agents.

At Stepnogorsk, the process of weaponizing anthrax would begin with a few grains of freeze-dried bacteria kept in a stoppered vial. Hundreds of tiny vials no bigger than test tubes were stored in metal trays inside a refrigerated vault, each over a soft towel soaked in disinfectant and each bearing a "passport" tag identifying the main features of the strain, including when it was created. One vial was enough to produce the munitions for an intercontinental war.

No one was ever allowed into the vault alone. At least two people—a lab technician and scientist—had to be present when a vial was taken down from the shelf, checked against a list, and wheeled in a metal cart into the operating laboratory.

We standardized the process after months of testing our production lines. First, the scientist would pour a small amount of a nutrient medium into the vial. The composition of this medium varies according to the strain being cultured, and the special formulas developed for what we called, with no attempt at irony, the "mother culture" were classified.

With a tiny pipette, the scientist would draw the mixture out of the vial and transfer a small amount into several slightly larger bottles. The bottles would be wheeled into another room, placed inside heated boxes about the size of a microwave oven, and left to incubate for one or two days.

Heat can kill bacteria, which is why pasteurizers of milk turn the temperatures of their ovens as high as 55 degrees Celsius to ensure that no harmful organisms remain in the product that goes to the supermarket. Weapons makers want their bacteria to survive, so one of the central challenges of bioweaponeering is to find the right temperatures at which different pathogenic microorganisms can grow rapidly without being cooked to death. This process has much in common with techniques for making vaccines.

A seed stock in a standard vial will swell to billions of microorganisms after less than forty-eight hours, but it would take days or even weeks of patient brewing to produce the quantities required for weaponization.

Once this liquid culture emerges from the thermostatic oven, it is siphoned off into large flasks. The flasks are brought to another room where they are connected to air-bubbling machines, which turn the liquid into a light froth. With oxygen distributed more evenly around the mixture, the bacteria can now grow more efficiently.

At this stage the liquid culture is translucent and deep brown, something like the color of Coca-Cola. The greater the bacterial concentration, the lighter and more opaque it will become: by the time it reaches maximum concentration, it will look like coffee mixed with cream.

A bioweaponeer works with recipes. The raw ingredients are similar, but quantities and combinations of nutrient media, heat, and time vary. If the mixture overheats, one has to begin the entire process over again.

Each new generation of bacteria is transferred into progressively larger vessels, until there is enough anthrax to pipe under vacuum pressure into a room containing several fermenters. These giant cauldrons incubate the substance for one or two more days. The bacteria continue to multiply until the scientist judges that they have reached maximum concentration, at which point they are passed through a centrifuge to be concentrated as much as thirty times further.

Our centrifuges resembled the separators used to make milk, butter, and cream at any dairy. In fact, they were produced for us by a plant in Tula, south of Moscow, that manufactures dairy equipment.

Even at this stage, you do not yet have a weapon. The pathogen has to be mixed with additives to stabilize it over a long period.

Like nutrient media, the additives are another "patented" element of the process.

The final formulation is sent along underground pipes to a nearby building, where it will be filtered into the munitions carrier. The machines that measure and pour fixed quantities of our pathogens dozens of times a minute are virtually identical to those used by soft-drink bottling plants. As the reactor is emptied, the seed stock from another tray of vials would have been cultured for the start of another cycle.

This process could be kept running night and day. Our experiments with different assembly-line techniques fueled our rapidly expanding program. By 1987, the combined production capacity of our anthrax lines around the country was nearly five thousand tons a year, although the actual mobilization plans authorized by the Ministry of Defense provided for a lower amount. Kurgan was to produce one thousand tons; Penza five hundred tons and Stepnogorsk three hundred tons.

My requests for equipment and building materials were rarely denied. My biggest problem was the shortage of staff.

About forty scientists were working at Stepnogorsk when I assumed control. Few of them were qualified for the advanced research that needed to be done. To fulfill Kalinin's vision, I would have to hire hundreds of new technicians and scientists, but the rules for employment at Stepnogorsk, as in all of our secret military installations, were rigorously enforced. Prospective workers had to undergo an intensive security check that could last months—months that I couldn't afford.

I knew Bulgak was right about the dangers of precipitately taking on staff, but the pressure to meet deadlines set by Moscow gave me little choice.

I launched an unofficial recruiting drive for construction workers, technicians, and scientists, tapping the work force in the city of Stepnogorsk and civilian institutes elsewhere in the country. Many of the people I brought to the base lacked proper clearances for secret work, so I hired them as temporary workers while their security checks were being done. Within a few months, I had almost two hundred new employees.

There were no awkward questions from headquarters, and our expansion and construction activities soon engaged so much of my time that I stopped worrying about Moscow's security rules. I recruited more people, quietly transferring them into full-time positions as soon as their personnel checks were completed. The success of our experiments, I imagined, had neutralized procedural concerns. But Anatoly Bulgak was not about to forget my insult. His network of informers kept him posted on my irregular hiring practices, and when he had accumulated enough evidence, he decided to teach me a lesson.

A year after our unfriendly encounter, I was ordered to Moscow. The order came directly from Kalinin. No explanation was offered, nor did I expect one. I assumed he wanted a personal report on our progress.

After a three-and-a-half-hour flight to Moscow on a cramped Aeroflot plane, I went straight to his office. Kalinin's secretary said he was busy. This didn't particularly surprise me. Kalinin was the type to order someone back from halfway around the world and then keep him waiting for days.

To my surprise, the secretary handed me a note from KGB colonel Vladimir Dorogov, counterintelligence chief for the entire Biopreparat organization.

"See me immediately," the note said.

Dorogov was staring out the window when I stepped into his office on the third floor, hands clasped behind his back. When he turned to face me, I was surprised by the ferocity of his expression.

"Do you realize how much danger you have put our country in?" he said coldly.

He walked to his desk and pulled out a folder with the names of the workers I had hired in the previous six months. There were red lines under several of them.

"We have excellent officers in Stepnogorsk," Dorogov continued. "But you seem to have chosen to resist their help. Frankly, I have never seen anything like this in my entire career."

His glacial calm was unnerving.

"Comrade Colonel, there is an explanation," I said.

"There can be no explanation!" he said. "I've seen your records and I know your history, Alibekov. This is not the first time you've been foolish."

He then gave me a blow-by-blow description of my encounter with Kuznetsov six years earlier.

"What shall we do about this?" he said.

"I don't know," I said, truthfully.

This was not merely a violation of procedure: in their minds I had opened the entire program up to sabotage. I began to believe that my career was rapidly drawing to an end.

But the KGB surprised me again. Dorogov opened the top drawer of his desk and pulled out a blank sheet of paper.

"On this sheet," he said, "you will write down everything you did and explain why it was wrong."

As I reached for it, he locked my wrist in a painful grip.

"Just remember," he said. "This is a small piece of paper, but it will have to cover your ass perfectly."

The next day Kalinin agreed to see me. He handed me the paper I had written, which was now attached to an official reprimand, and told me to sign my name to both sheets. It was a tradition that went back to Stalin's time: my signature affirmed the charges against me and concurred in advance with whatever "people's punishment" I was deemed to deserve.

My humiliation, apparently, was my punishment.

"If you ever break the rules again," Kalinin said, "it will be the last time."

Today, I understand that they could not have fired me, but it was not obvious to me then. Putting another manager in my place would have slowed the momentum of our program, perhaps even derailed it, and Kalinin had too much invested in Stepnogorsk's success. He must have used all of his influence to fight off the KGB. If I failed, his career would be destroyed as surely as mine.

I shared Kalinin's determination to succeed at any cost, which added irony to my predicament. The idealistic young doctor from Tomsk who had agonized over the difference between saving lives and taking them was gone. The worst possible fate for me had become banishment from Biopreparat, and from the privileges that came with it.

The transformation wasn't yet complete. I still shuddered occasionally when I looked at the bacteria multiplying in our fermenters and considered that they could end the lives of millions of people. But the secret culture of our labs had changed my outlook. My parents would not have recognized the man I had become.

I returned to Stepnogorsk determined to work with the dispassionate efficiency Biopreparat required of all its managers. My family took second place. Some weeks I lost all track of time while I was inside the compound, going home only for a nap and a snack before setting off again. In 1985, my third child, Timur, was born, but I was almost never there: while Lena took care of the baby, I was working feverishly in the labs.

I was angrier and lonelier than I had ever been in my life. When I first arrived at Progress, I had been filled with excitement about returning to Kazakhstan after so many years in the Russian north. Stepnogorsk was only a short plane ride from my parents' home, and I looked forward to being around people who looked like me and who spoke the language I had learned as a child. But there were no Kazakhs at my facility, and only a few were scattered among the Russian faces in the city of Stepnogorsk. Even as I had become the perfect model of a Soviet bureaucrat, I felt alienated from everything and everyone around me.

My oldest child, Mira, was treated well by her teachers and friends because she was the daughter of the director, but I knew that some of her classmates mocked her as "black" and called her "funny-face" behind her back.

The few quiet moments I stole in our apartment were used to work on my Ph.D. thesis. It was absolutely necessary to complete the thesis if I was to maintain my career path at Biopreparat. Nothing else seemed important.

Eventually, Bulgak and I came to an uneasy truce. He was shrewd enough never to ask me about my trip to Moscow, although he clearly relished my humiliation. He would eventually, to my relief, be transferred back to his provincial detachment, but not before he became embroiled in a security problem of his own.

One of the units under Bulgak's control was called the Division for Special Countermeasures Against Foreign Engineering Intelligence Services. It was a convoluted title for the straightforward job of making sure that nothing we did at Stepnogorsk was detectable in the outside world.

My improvements to the plant had complicated Bulgak's life. He needed people with sufficient technical expertise to mask all traces of the prodigious flow of waste from our fermenters. Bulgak found a civilian engineer to head the Countermeasures Division who soon proved to be one of our most talented workers in the arts of camouflage. His name was Markin.

Markin was a shy man in his late thirties or early forties. Although most of his coworkers liked him, he kept largely to himself. Few knew how complicated his personal life was.

Markin had fallen in love with the widow of one of the KGB officers who had worked at the plant. They married after a brief courtship, but the marriage quickly soured. They fought constantly, and Markin began to look more downcast with each passing month.

Finally, he applied for a leave of absence, explaining that he needed to take care of his sick mother. The leave was granted. A few weeks later, Bulgak walked into my office holding a letter from a small village in the Gorky region.

"Read this," he said, his brow furrowing with anxiety.

Markin wrote that he didn't want to return to Stepnogorsk. "I respectfully ask that you allow me to retire to the collective farm where my ailing mother lives," the letter said. "But I beg you not to think that there is any other reason for my departure. I am not a traitor, just an insignificant person who would like to live in the peace of nature."

I handed the letter back to Bulgak.

"I guess he's found a way to escape his wife." I smiled.

Bulgak didn't smile. "We can't let him go," he insisted. "He knows too much."

"He knows a lot," I agreed. "But what foreign spy is going to plod through the mud of Gorky to find him out? I don't think we have anything to worry about. Besides, he's not in the army or the KGB. You can't keep him here."

Bulgak looked away distractedly. "We'll see," he said.

A few days later, I found Bulgak in a more cheerful mood.

"One of my men just spoke with the local commander in Gorky about the Markin problem," he grinned. "The guys there are complaining that they have two headaches now."

"What does that mean?"

Bulgak gave me a pitying look. "If you followed politics," he lectured, "you would learn what's really going on in the world. Don't you know that's where they've got Sakharov?"

Andrei Sakharov, a Nobel Prize–winning physicist and the father of our hydrogen bomb, had been exiled to Gorky in 1980 for public criticism of the Soviet leadership. It seemed strange to put Markin in the same class of "headache" as the outspoken physicist.

Bulgak and I were going over new security regulations some weeks later when he sat up in his chair like a man pricked by a needle.

"What's wrong?" I asked.

"I almost forgot," he said. "Remember the two headaches I told you about? In Gorky?"

I told him I did.

"Well," he went on, relishing every word, "Gorky only has one headache again."

"What does that mean?"

"It means," said Bulgak, "that Markin is no more."

"You mean he left Gorky?"

"Unwillingly," he said. "He's dead."

"What happened?" I asked uneasily.

"It seems he drowned. He was drinking a little too much and he went out for a swim and never came back."

"I didn't know Markin liked to swim."

An enigmatic smile played on Bulgak's features.

"The important thing is that Gorky has only one headache again," he said.

"Was he killed?"

Bulgak looked hurt.

"How would I know?" he said. "What matters is that we don't have to worry about Markin anymore."

———

By 1986 we had over nine hundred people at the plant, and more were coming every month. The contingent from Sverdlovsk, which included the unfortunate Nikolai Chernyshov, helped us achieve a breakthrough in developing the most effective anthrax weapon ever produced. But the pressure-cooker atmosphere took a heavy toll. There were one or two accidents every week.

Once Gennady Lepyoshkin, the chief of our biosafety directorate, reported that a technician had been infected with anthrax in a lab that was supposed to be sterile. He had an abrasion on his neck, one of the most dangerous places in the body through which to contract cutaneous anthrax. When the neck swells, it interferes with breathing.

At first we treated him with streptomycin and penicillin, the most effective antibiotics for use against cutaneous anthrax, but a painful swelling erupted on his chest and spread over his body, making it increasingly difficult for him to breathe. Within three days, death seemed inevitable. A gloomy message was being prepared for Moscow when, in a final attempt to save his life, we gave him an abnormally high dose of anthrax antiserum. The shock dose worked: he began to recover.

The technician's narrow escape drove home the potency of our new weapon. Our powdered and liquid formulations of anthrax were three times as strong as the weapons that had been manufactured at Sverdlovsk. It would take only five kilograms of the Anthrax 836 developed at the Kazakhstan base to infect half the people living in a square kilometer of territory; the Sverdlovsk weapon needed at least fifteen kilograms to achieve the same impact.

The destructive power of the new weapon was confirmed in tests on Rebirth Island in 1987. Lepyoshkin, who became my senior deputy that year, flew down to the Aral Sea to supervise the field trials. When he reported success, Moscow finally took Sverdlovsk Compound 19 off the roster of anthrax production plants.

Stepnogorsk more than compensated for the lost capacity of the army plant. Our factory could turn out two tons of anthrax a day in a process as reliable and efficient as producing tanks, trucks, cars, or Coca-Cola.

With the creation of the world's first industrial-scale biological weapons factory, the Soviet Union became the world's first—and only—biological superpower.

To be sure, we had already achieved global dominance in this field by the 1980s, when we could launch a biological attack with intercontinental ballistic missiles on targets thousands of miles away. But Stepnogorsk demonstrated our ability to wage biological warfare on a scale matched by no other nation in history. We had taken the science of biowarfare further in the previous four years than it had traveled in the four decades since World War II.

Needless to say, we didn't advertise our accomplishment. The accident at Sverdlovsk had briefly opened a window on our biological warfare program to the outside world, but since then our secrets had stayed well hidden. The international community still knew nothing of Biopreparat, and it had no reason to suspect our program's growing dimensions.

9

SMALLPOX

Moscow, 1987

The word *virus* comes from the Latin term for poison. Viruses are invisible under most microscopes and hundreds of thousands of times smaller than a grain of sand. Their existence was unsuspected until Dmitry Ivanovsky, a Russian microbiologist, discovered them in the late nineteenth century while investigating an outbreak of mosaic disease in tobacco plants. Ivanovsky found that the mysterious agent responsible for this disease was able to pass through filters that otherwise blocked bacteria. Over half a century would pass before the first virus was seen and identified under an electron microscope, but Ivanovsky's discovery launched a new field of research into infectious diseases.

As more viruses were discovered, scientists grew increasingly baffled by their behavior. Viruses seem to exist on the threshold of life, remaining inert until they fasten onto the cells of other organisms. They are structurally simpler than bacteria, consisting only of a protein shell, a sequence of DNA or RNA, and sometimes a lipid membrane, but they are capable of annihilating the most sophisticated biological system. Not all viruses kill their hosts—to do so is in many ways impractical—but when they do, they often combine incredible virulence with a high degree of contagiousness. A virus is programmed for its own procreation, but it cannot do this alone. First it must locate a host with the cell structure and nutrients necessary for it to reproduce. Viruses come to life inside the nucleus or cytoplasm of their host cells, fusing with them and ultimately hijacking their functions.

The human body commands a number of complex mechanisms for resisting, containing, and killing pathogenic microorganisms. The immune system works on many levels at once, like an army with scouts and infantrymen, naval and air power, a sophisticated information network, and a carefully delineated command structure. Some cells are responsible for surveillance, others for coordinating information; some focus on local maneuvers while still others direct more general attacks. Immunologists distinguish between specific and nonspecific immunological reactions. Specific or acquired immune responses depend on memory cells, which store information about previous invaders and thus play a significant role in conferring immunity.

Among the most important agents in the immune system are T cells. They act as scouts, circulating through the bloodstream and moving into lymph nodes, on the lookout for foreign substances. As soon as a virus enters the bloodstream and infects its first cell, it will be recognized by T cells, which immediately activate, replicating themselves and sending out signals, calling for the formation of antibodies and attracting them to the site of infection. Antibodies are like ground troops. They are particularly effective at attacking viruses and bacteria that are still coursing through the system, before they have infiltrated target cells.

Within seconds of infection, defensive proteins and inflammatory agents are released, which activate natural killer cells and lead them to the site of infection. Interferon, one of the most powerful antiviral agents, degrades viral RNA, slows down protein synthesis, and inhibits viral reproduction in infected cells.

By the end of the first week or the beginning of the second, the body will in many cases have developed virus-specific antibodies, which sometimes seek to neutralize the virus by binding to its surface and preventing it from penetrating into new cells. But viruses are adept and mutate quickly. Countless are now capable of inhibiting and neutralizing the body's natural defenses, rendering their resistance ineffective.

New viruses can appear without warning, and viruses once considered harmless to man can suddenly morph into killers. They can be responsible for devastating epidemics, such as AIDS or Ebola, or they can be benign, like the virus that causes some warts. Some viruses only infect plants. Others target animal life. Arboviruses, transmitted by insects, usually aim for the brain, muscles, liver, heart, and kidneys. Enteric viruses lay siege to the gastointestinal tract, entering the body through contaminated water or food. Respiratory viruses, responsible for measles, mumps, and chicken pox, are airborne viruses that assault the nose and throat. More than one hundred different viruses have been identified as causes of the common cold.

Of all the diseases that have tormented mankind, smallpox has left the oldest and the deepest scars. Recorded as early as 1122 B.C. in China, it altered the course of history, ravaging eighteenth-century Europe and decimating the native populations of North America.

Smallpox comes from the pox family of viruses, which assault the upper respiratory tract. *Variola major,* the scientific name by which the smallpox virus is known, is patient and systematic. It will begin by insinuating itself into cells close to the surface of the skin and in the neural system. The smallpox virus sheds its shell as soon as it enters a live cell, and quickly begins to multiply. Viral transcription begins almost immediately, inhibiting DNA synthesis and thereby preventing the cell from activating its defense mechanisms. Once the virus has inserted its genetic information into the host cell, proteins and enzymes are created to help it mature and develop. The progress of the virus can be mapped by the spread of tiny pink spots from the face and arms to the lower regions of the body.

Smallpox symptoms were once familiar to every doctor. After a quiet incubation period of five to ten days, the virus manifests itself suddenly. The first stage of the disease brings high fever, vomiting, headache, and a strange stiffness. This can last from two to four days. Within less than a week, small spots will begin to develop, forming a rash around the face. As the rash spreads over the following week these spots will develop into painful blisters. In the normal course of the illness, the blisters form scabs that linger for several weeks until they dry and fall off, leaving scars. More severe forms of black or red pox can lead to death within three to four days.

The modern struggle to conquer smallpox began in 1796, when the British physician Edward Jenner observed that milkmaids who had contracted a mild form of pox virus from cows appeared to be immune to smallpox. Jenner injected an eight-year-old boy with material taken from lesions on the hand of an infected milkmaid. The boy developed a slight fever. Two months later Jenner inoculated him with smallpox, but he didn't contract the disease. The physician concluded that the milder strain, which he named *vaccinia,* provided immunity.

Smallpox "vaccine"—the name chosen to honor Jenner's work—became the principal instrument for tackling the disease. His discovery, the first vaccine, revolutionized medicine.

On May 8, 1980, the World Health Organization announced that smallpox had been eradicated from the planet. The last naturally occurring case was reported in Somalia in 1977, and no new cases had been detected in three years. The WHO recommended the discontinuation of smallpox immunization programs, observing that there was no longer any need to subject people to even the negligible risk connected with vaccination.

The international agency simultaneously adopted a resolution restricting the world's stocks of smallpox to four sites, where limited quantities would be available for research purposes. A few years later, the sites were narrowed down to two: the Centers for Disease Control in Atlanta and the Ivanovsky Institute of Virology in Moscow.

The conquest of smallpox generated a special feeling of accomplishment in the Soviet Union: the worldwide crusade against smallpox had been a Soviet initiative. Moscow first proposed the campaign at a World Health Organization meeting in 1958, and its sponsorship of vaccination programs in the third world won it admirers everywhere. Russia had suffered its share of smallpox outbreaks over the centuries, finally managing to eliminate the disease in 1936, after a decade-long immunization program sponsored by the fledgling Bolshevik government.

Soon after the WHO announcement, smallpox was included in a list of viral and bacterial weapons targeted for improvement in the 1981–85 Five-Year-Plan.

Where other governments saw a medical victory, the Kremlin perceived a military opportunity. A world no longer protected from smallpox was a world newly vulnerable to the disease. In 1981, Soviet researchers began to explore what the Kremlin hoped would be a better version of a smallpox weapon that had been in our arsenal for decades. The work was at first cursory. Military commanders were reluctant to devote energy and resources to an enterprise that promised no immediate results. The Soviet Union, they reasoned, had already gone further with smallpox weapons than any other country.

In 1947, the Soviet Union established its first smallpox weapons factory just outside the ancient cathedral town of Zagorsk, forty minutes' drive northwest of Moscow. Zagorsk (now Sergiyev Posad) is the site of the walled fourteenth-century Trinity–St. Sergius Monastery, one of the most revered places in the Russian Orthodox religion. A few miles away, in another walled compound, Soviet army scientists at the Virological Center of the Ministry of Defense devoutly cultivated smallpox, Q fever, and Venezuelan equine encephalitis in the embryos of chicken eggs.

It was a cumbersome process, but an effective one. Using tiny syringes, laboratory workers injected microscopic amounts of smallpox virus into eggs and sealed each egg with paraffin. The eggs were placed inside thermostatic ovens for several days while the embryo host cells stirred the virus into life. As it monopolized the cells' normal growth mechanisms, the virus spawned successive replications of itself until the host was engulfed or destroyed. The eggs were then punctured and the liquid inside poured into special vats and mixed with stabilizing materials. The resulting weapon could remain potent in refrigerated conditions for at least a year.

Every month, hundreds of thousands of eggs produced at nearby collective farms were consigned to Zagorsk's weapons-assembly lines. Under a state-controlled agricultural system it was easy to conceal the purpose of hijacking so many eggs from the marketplace. The "egg-weapon" process for smallpox proved so successful that a second production facility was opened at Pokrov, near Moscow, in a plant operated by the Ministry of Agriculture.

In 1959, a traveler from India infected forty-six Muscovites with smallpox before authorities realized what had happened. The traveler had been vaccinated, but smallpox vaccinations lose their effectiveness over time, and while his weakened immunity was enough to protect him from suffering the symptoms of the disease, he could still pass it on to others. The strain of *Variola major* in his system was so virulent that an epidemic was only narrowly averted. Partly in response to this incident, the Soviet government sent a special medical team to India to help purge the virus from the subcontinent.

KGB agents went with them.

They returned to Russia with a strain of Indian smallpox excellently suited to weapons production. It was highly virulent and was stable enough to retain its infectious qualities over time. This meant that, with the proper additives, it could be stored longer than the strains in Soviet stockpiles. Within a few years, India's unwitting gift became our principal battle strain of smallpox. It was dubbed India-1967, to commemorate the year of its isolation. In our secret code, it became India-1.

In the 1970s, smallpox was considered so important to our biological arsenal that the Soviet military command issued an order to maintain an annual stockpile of twenty tons. The weapons were stored at army facilities in Zagorsk. Annual quotas of smallpox were required as it decayed over time. We never wanted to be caught short.

The episode of the Indian traveler underlined some of *Variola major*'s impressive qualities as a weapon. The smallpox virus is so hardy that it can remain infectious for long periods, even in the soiled linen of those who have been infected. Smallpox victims are infectious from the moment of their first symptoms until the healing of the last scar, two to three weeks later, and they can transmit the disease to others with as little as a cough.

Not all viruses spread through direct human contact, but those that do are the most contagious. Influenza's ability to travel from one person to another propelled by a sneeze or a cough or even by touching the clothes of an infected person enables it to leapfrog through classrooms and cross international borders and to fuel lethal pandemics such as the flu of 1918. Measles and chicken pox, the tormentors of early childhood, are similarly spread through ordinary human contact, which means that the virus can remain alive outside its host long enough to travel through the air and infect another person. Some of the viruses for hemorrhagic fevers—Machupo, Ebola, Lassa fever, and Junin—spread through personal contact. Ebola ravages its hosts and dies quickly on exposure to oxygen, though a number of people who have had only glancing contact with Ebola victims have come down with the disease.

A large body of Western scientific opinion considers *Variola major* an unlikely weapon, despite its contagiousness. Smallpox does not normally occur in animals, though monkeys can contract it without spreading it to others. Humans are the virus's only natural hosts. There is no way, therefore, for the disease to propagate in nature. Some scientists argue that an outbreak among humans would be contained through quarantine and vaccination before it could develop into an epidemic, as happened in Moscow in 1959.

These scientists go on to maintain that the virus's long incubation period offers additional protection. Since the disease takes seven to ten days to incubate, they claim there would be plenty of time to take medical countermeasures after a smallpox attack. The first order of priority would be to contain the outbreak through a program of blanket vaccination. Smallpox vaccines are thought to become effective within a few days. They were found in clinical tests performed many years ago to reduce the severity of the disease. But there is no guarantee that vaccination will cure smallpox. And in order to be effective, they must be administered before the first symptoms appear.

The smallpox weapons we developed sharply reduced this comfort period. When we exposed monkeys to an aerosol of the highly virulent India-1, they contracted smallpox within one to five days.

There are no therapeutic measures currently available to treat smallpox once symptoms develop. The most a physician can do at this stage is to provide drugs to alleviate some of the symptoms.

A virus's effectiveness as a weapon can be measured by its morbidity rate, which reflects the number of people to contract the disease after exposure. Smallpox kills between 30 and 50 percent of unvaccinated victims, a low mortality rate, but its morbidity rate ranges from 60 to 90 percent. For many people, contracting smallpox amounts to a life sentence. Some victims are permanently blinded. Others will bear scars as long as they live.

In the nearly twenty years since it issued its declaration, the World Health Organization has not modified its stance on smallpox. Schoolchildren in the United States, Russia, and elsewhere around the world are not vaccinated against the disease, and international travelers are no longer required to show proof of smallpox immunity.

Today there are twelve million doses of smallpox vaccine on hand in the United States—of which only seven million are fully reliable, according to the Centers for Disease Control in Atlanta—a portion of the roughly two hundred million doses available in the world. This sounds like a comfortable amount to meet an emergency, until you consider the damage a smallpox attack would do in a densely populated commuter city like New York.

Viruses have attracted the attention of bioweaponeers for decades. During World War II the Western allies explored the possibility of weaponizing several viral diseases, including Venezuelan equine encephalitis and smallpox. American, Canadian, and British scientists found to their frustration that viruses were far more difficult to manipulate than bacteria. Since they cannot grow on their own, they must be nurtured in living cells or tissues inside a sterile laboratory environment.

Viruses were also found to be unreliable when deployed. Aerosols were still in the early stages of development in the 1940s, and most of the approaches considered by the Allies for weaponizing smallpox seem strange today. One method involved grinding an Asian strain of smallpox into a fine powder to dust over letters. By the time the war ended, the Allies had largely given up on weaponizing viruses.

The Soviet Union was not deterred. Throughout the Cold War, we considered viruses to be among the most valuable munitions in our arsenal. Their ability to infect vast numbers of people with an infinitesimal number of particles made them ideal weapons for modern strategic warfare. As our technical ability to create aerosols improved, we found they could be used to greater effect than some bacteriological munitions, especially in the case of diseases spread through direct personal contact. Fewer than five viral particles of smallpox were sufficient to infect 50 percent of the animals exposed to aerosols in our testing labs. To infect the same percentage of humans with anthrax would require ten thousand to twenty thousand spores. For plague, the comparable figure is fifteen hundred cells. The differences in quantity are too minute to be discernible to the naked eye, but they are significant if you are planning attacks on a large scale. Smallpox requires almost no concentration process.

While we had clung to our egg-and-conveyor-belt method of making smallpox, Western pharmaceutical labs were manufacturing vaccines in special reactors from cultures grown in tissue cells obtained from animals or humans. This technique required expertise. The tissue has to be kept alive outside its natural habitat in cell lines and stored at precise temperatures. Some cells are better than others for viral culture, such as those taken from the kidneys of green monkeys or from the lungs of human embryos.

The nutrient media needed to cultivate tissue cultures are unlike those used to grow bacteria. A special complex of amino acids, vitamins, salts, and sera—all mixed with highly distilled and deionized water—is crucial to the processes that encourage tissue cells and ultimately viruses to grow.

The new methods being developed were far more efficient than ours, and much easier to conceal.

I was promoted to colonel for my work in Stepnogorsk. This was two years ahead of the normal military schedule, but the real reward, as far as my family was concerned, came when General

Kalinin transferred me to Moscow in September 1987 to serve as deputy chief of the Biosafety Directorate at Biopreparat. He hinted I was in line for higher posts.

The transfer surprised me. Despite my achievements, Kalinin had begun to treat me with hostility. He was curt on the phone and disparaged my work to others in the program. He had even opposed my receiving a medal for the new anthrax weapon.

"We're giving Alibekov something every year," he complained to a colleague who later relayed the comment to me. "He's too young to move so fast."

Fortunately, I had powerful allies. Senior officials in the civilian hierarchy of the Military-Industrial Commission (VPK) and the Fifteenth Directorate saw my appointment to headquarters as an opportunity to exert leverage over Kalinin.

"You remind me of myself," Alexei Arzhakov, deputy chairman of the VPK, confided to me at one point during the tug-of-war that preceded my transfer. "I became director of a chemical weapons production facility at thirty-three."

But it was really Mikhail Gorbachev and his revolutionary restructuring program, *perestroika,* that sent me to Moscow. Gorbachev came into office in March 1986, determined to break up the corrupt bureaucratic fiefdoms of the Brezhnev era and to create a stronger, more cohesive government. He was the reformer my generation had been waiting for. Nearly everyone under forty at Biopreparat considered him our best hope.

The biological warfare establishment was an ideal candidate for reform. By the mid-1980s, it was a jumble of agencies, laboratories, and institutes constantly trying to undercut one another. The enterprise languished under the control of bureaucrats as sclerotic as those who were stifling the Soviet Union's progress in other areas.

Perestroika produced encouraging early results in our program. Biopreparat and Glavmikrobioprom, responsible for producing vaccines and medicine, were placed under the control of a new superministry, the Ministry of Medical and Microbiological Industries. Valery Bykov, a veteran apparatchik and a specialist in chemical warfare, became minister. Yury Kalinin was named deputy minister.

Their combined leadership seemed like a recipe for disaster. The two men were old rivals who had sparred for control of biological warfare research during the Brezhnev and Andropov years. The quarrel was as much institutional as personal. From its inception, Biopreparat was at the center of a wrestling match between the army and civilian officials. Brezhnev had granted the Fifteenth Directorate almost complete freedom to set Biopreparat's budget and its research program, and to choose its personnel.

Under Gorbachev, the army's honeymoon came to an end. Soviet military commanders found themselves challenged by the architects of *perestroika* in nearly every sphere. At Biopreparat Kalinin was forced to accept greater civilian control. When Bykov decided to support my transfer to Moscow, my future was assured.

Kalinin managed to turn the situation to his advantage. Reversing his position, he soon became my greatest supporter, making it seem as if transferring me to Moscow had been his idea all along. He had his own version of *perestroika* in mind.

Kalinin would use me to unseat his rivals—General Lev Klyucherov, head of the scientific directorate, and General Anatoly Vorobyov, a dignified elderly scientist whom I would replace as first deputy director within the year.

In December 1987, three months after I arrived in Moscow, Kalinin presented me with my first big assignment: I was to supervise plans to create a new smallpox weapon.

I spent an afternoon inside the third-floor KGB archives at Samokatnaya Street reading my instructions, contained in a secret document setting out the goals of Soviet biological weapons development for the five-year period ending in 1990. Smallpox appeared as a "special item" among the list of diseases marked for weaponization.

The Five-Year Plan, signed in his characteristic scrawl by Mikhail Gorbachev, outlined the most ambitious program for biological weapons development ever given to our agency. It included a three-hundred-million-ruble viral production plant (then equivalent to four hundred million dollars) at Yoshkar-Ola in the autonomous republic of Mordovia. The plan established a new military facility at Strizhi, near Kirov, for the production of viral and bacterial weapons and, most significantly, it funded the construction of a 630-liter viral reactor to produce smallpox at the Russian State Research Center of Virology and Biotechnology, a facility known within The System as Vector. Our military leaders had decided to concentrate on one of the toughest challenges of bioweaponeering—the transformation of viruses into weapons of war.

Gorbachev's Five-Year Plan—and his generous funding, which would amount to over $1 billion by the end of the decade—allowed us to catch up with and then surpass Western technology.

When I went to Vector in 1987, our new smallpox project was just getting off the ground. The facility, founded by Biopreparat in the early 1970s to specialize in viral research, was located in the small Siberian town of Koltsovo. It had been left to stagnate while we focused on improving our bacterial weapons, but Gorbachev's decree gave it a new lease on life.

Dozens of new lab and production buildings earmarked for research into viruses had been constructed by the time I arrived. More were on the drawing board. There was a large biocontainment structure designed especially for laboratory experiments with contagious viruses such as smallpox, Marburg, Lassa fever, and Machupo, as well as new explosive test chambers and facilities for breeding animals.

Vector's prize acquisition was the expensive new viral reactor authorized by Gorbachev's decree. Designed by one of our Moscow institutes and assembled at a special Biopreparat plant in western Russia, it was the first of its kind in the world. It stood about five feet high and was enclosed within thick stainless steel walls. An agitator at the bottom kept the mixture inside churning like clothes in a washing machine. Pipes led out in several directions, both for waste matter and weapons-ready material. A window on its convex roof allowed scientists to observe the viral culture at all times.

Lev Sandakchiev, Vector's director, was a garrulous Armenian biochemist who had been with Biopreparat since its inception in 1973. Sandakchiev was an expert in orthopoxviruses, the viral genus that includes smallpox. When I saw him he was at his wits' end.

New scientists and technicians were arriving at Vector every month as the program began to take shape. He had to arrange their housing and set up their work programs while keeping track of the construction projects. The scholarly virologist had been leading a backwater scientific research group of several hundred people; now he found himself supervising a work force of more than four thousand workers.

"Just tell me what you need, and I'll get it for you," I said, determined to meet what I regarded as the first test of my talents as a senior executive.

Sandakchiev gave me a haughty look, as if I were one of his lab assistants.

"Time," he replied. "Can you give me time?"

I may have impressed the military and the bureaucrats in Moscow, but there was widespread skepticism inside our elite scientific community about my qualifications for the job. I felt this implicit criticism in his pointed remark.

"I can't give you time," I said with a smile. "It's the one resource we're not permitted to exploit."

As the months progressed, Sandakchiev and I developed a respectful working relationship, and I was able to unsnarl some of the bureaucratic logjams that had been making his life impossible. At first our biggest concern was safety. If even a tiny amount of smallpox were to escape into the surrounding countryside, it would cause a horrific epidemic. It would be much harder to cover up than the anthrax outbreak in Sverdlovsk.

Sandakchiev was determined to protect his employees. He repeated time and again that he would not sacrifice the health of a single worker to the pressure of a deadline. But running a biological weapons plant was not like managing a small research facility. New rules had to be enforced, and there were higher expectations. To keep the country—and our program—safe from exposure, Moscow imposed quarantine conditions on all Vector employees engaged in smallpox research. The staff was confined to special dormitories near the compound and guarded around the clock by security police. In a compromise, we granted them periodic leave to visit their families.

Considering that outsiders might be suspicious if they saw hundreds of people with the distinctive marks of fresh smallpox inoculations on their arms years after the Soviet Union had discontinued all immunization, we decided, after some deliberation, to issue a directive that workers be inoculated on their buttocks. We assumed this part of their anatomy was safe from prying foreign eyes.

Despite his laboratory expertise, Sandakchiev knew little about the technological process required to mass-produce smallpox. We needed someone who was not only a smallpox expert but who could make our new equipment and production lines work efficiently. A search of Biopreparat's personnel records turned up no one in the country who satisfied both requirements. Without such a production manager, the project was sure to falter.

I was at my desk early one morning in Moscow when Sandakchiev's excited voice came through on the phone.

"I've found the man we need," he said. "But I'm going to need your help to get him here."

I recalled with trepidation the trouble I'd gotten into at Stepnogorsk for my unorthodox hiring policies.

"I'll do my best," I said cautiously. "Who is he?"

"His name is Yevgeny Lukin. He's a colonel, works for the Fifteenth Directorate at Zagorsk. No one in the country knows more about producing smallpox. I've already spoken with him and he wants to come. We need you to do the paperwork."

I hadn't thought of the Fifteenth Directorate. The army command's jealousy and its distrust of Kalinin made personnel transfers between the directorate and Biopreparat almost impossible to arrange.

I made a few calls. Sandakchiev was right: Lukin was perfect for the job. As a young scientist at Zagorsk in the 1960s, he had been one of the luminaries of the early smallpox weaponization program. I decided to invite him to Moscow for an interview with Kalinin.

Lukin was in his early fifties but he carried himself with the military bearing of a younger man. I liked him at once.

The interview was excruciating. Kalinin fired questions relentlessly and with each passing moment Lukin seemed to sink deeper into the floor.

"Yevgeny," the general drawled, "I don't remember ever hearing you stutter before. Is this a new defect?"

Suddenly, I remembered that Kalinin had spent part of his early career at Zagorsk. The two men were almost contemporaries. They obviously knew each other. Whatever relationship they had once enjoyed, Kalinin was determined not to let him forget the difference in their status.

The interview over, the terrified colonel was finally permitted to leave. I was about to follow when Kalinin motioned for me to stay behind. He had evidently enjoyed himself.

"He's not a bad guy," he said. "I don't see why he was so frightened."

"A lot of people are scared of you," I said.

Kalinin bent his head over his desk. I couldn't see the expression on his face, but I suspected my comment pleased him.

"All right," he said finally. "Sign the order and make him deputy at Vector."

Any doubts I had had about Kalinin's ability to manage the transfer soon vanished: Lukin was on his way to Siberia within a week.

From that point on, my opinion of Kalinin began to change. He had dominated my life almost from the moment I had joined Biopreparat. Like many of my colleagues, I resented his manipulative behavior and cool arrogance. But we all understood that those traits had helped secure the organization's place in our cutthroat political world. At closer range, however, they were even less attractive. I knew I owed my status to him, but as I watched him deal with his subordinates every day in the same callous manner in which he had treated Lukin, I wondered whether some day he would, at a whim, crush me too.

Lukin's transfer was one of the best decisions Biopreparat ever made. Lukin was able to create a production line to manufacture smallpox on an industrial scale, and over the next year I watched with growing satisfaction as Vector blossomed under Sandakchiev's management into a formidable weapons development complex.

In December 1990, we tested a new smallpox weapon in aerosol form inside Vector's explosive chambers. It performed well. We calculated that the production line in the newly constructed Building 15 at Koltsovo was capable of manufacturing between eighty and one hundred tons of smallpox a year. Parallel to this, a group of arrogant young scientists at Vector were developing genetically altered strains of smallpox, which we soon hoped to include in this production process.

10

VECTOR

Koltsovo, Siberia, 1988

The windows in the administrative offices at Vector were covered with thick sheets of ice. It was midway through the Siberian winter, and the temperature outside had plunged to minus forty degrees Celsius. The scientists crowding into the tiny room were bundled in sweaters and thick jackets. They grumbled about the cold and the peculiarities of the Soviet food-supply system.

"I don't remember the last time I saw a fresh tomato, or an orange," one called out.

"We're going to have to start stealing from our animal cages," said another, to a burst of laughter.

I smiled good-naturedly. It was February 1988, and I was on one of my frequent commuting trips to the Vector institute. By then I knew the scientists well enough to enjoy their bleak sense of humor.

The man whose joke provoked so much laughter was a hardy example of our Siberian species of scientists. His name was Nikolai Ustinov. A gregarious, well-built man with an easy smile and a sharp wit, Ustinov led a research team working on Marburg, a hemorrhagic fever virus we had obtained in the 1970s. Marburg was set to become one of the most effective weapons in our biological arsenal. The project had become as important as our work with smallpox.

Ustinov loved his job. He had been at Vector for many years and was one of the most well liked members of the community. He enjoyed socializing after hours with his colleagues almost as much as spending time in the lab. His wife, Yevgenia, worked as a lab scientist in another part of the institute, and the couple had two teenage sons. He was forty-four when I met him.

Before we settled down to discuss the serious business of the morning, I made a mental note to ask Ustinov if there was anything I could do to improve the food situation. Unfortunately, I forgot to ask him.

Two months later, in mid-April, I was sitting in my Moscow office one morning when a call came in from Lev Sandakchiev, Ustinov's boss and the head of Vector.

"Something terrible has happened," he said.

"An accident?"

"Yes. It's Ustinov. He injected Marburg into his thumb." Sadness and frustration were palpable in his voice.

"Right into his thumb," he repeated. "He was in the lab working with guinea pigs when it happened."

"Wait," I interrupted him. "You know the regulations. Send me a cryptogram. Don't say any more."

I felt heartless ordering Sandakchiev to stop talking, but the mere mention of Marburg was too sensitive for an open line.

Marburg was the most dangerous virus we were working with at that time—dangerous because we knew so little about it as well as because of its terrible impact on humans.

The first recorded outbreak of the virus occurred in 1967 at the Behring pharmaceutical works in Marburg, an old university town seventy miles north of Frankfurt. An animal keeper died two weeks after he contracted a mysterious illness from green monkeys sent to the Behring lab from central Africa. The lab was culturing vaccines in kidney cells extracted from the monkeys. Other workers soon fell sick, and similar cases were reported at laboratories in Frankfurt and Belgrade, both of which had received shiploads of green monkeys from central Africa at the same time.

Twenty-four lab technicians came down with the unknown disease, along with six of the nurses caring for them. Of the thirty-one people infected, seven died. This kind of undiagnosed outbreak would be alarming enough, but it was the horror of their deaths that caught the attention of biologists and tropical disease specialists around the world.

The mysterious virus appeared to liquefy body organs. One of the survivors went mad after the organism chewed away his brain cells. Before the victims died, every inch of their bodies was wet with blood.

Following tradition, the virus was named after the place where it was first identified. It would alter forever the image of a city that has been a center of European philosophy, science, and religion for centuries.

Some of the world's greatest bacteriologists and biochemists have studied at Marburg—including Albrecht Kossel, whose research laid the groundwork for the discovery of DNA, and Alexandre Yersin, a codiscoverer of the plague bacterium (named *Yersinia pestis* after him). The lab in which Marburg was first smeared on a glass slide was itself named after the man credited with founding the science of immunology—Emil von Behring.

A similar virus surfaced nine years later on the banks of the Ebola River in Zaire, now the Democratic Republic of Congo. By the time that epidemic died out, 430 people were dead in Zaire and nearby Sudan. The virus responsible for that outbreak was called Ebola, after the site where it was isolated. Ebola struck again in the same area in 1995.

The viruses isolated in Africa differed slightly in genetic composition from the strain found in Germany, but they were closely related. Under an electron microscope, both organisms seemed to proliferate by shooting out tiny filament-like threads, like the lines cast by fishermen, from the cells they had already scoured for the food they needed to grow. The threads were often bent at the top, like fishing hooks, and as they prepared to invade a new cell they curled into rings, like microscopic Cheerios. Marburg and Ebola were deemed to belong to a new family of viral organisms. They were called filoviruses.

We still know very little about where filoviruses come from and how they are transmitted to humans. In some cases an animal or insect bite has delivered the organism into the bloodstream. In others, sexual contact has been a source of infection, and some scientists believe the virus may even be located in plants. Both Ebola and Marburg can spread from one person to another with no direct physical contact. Some victims in Germany and in Africa had merely been in the same room with infected patients. Ebola's mortality rate is between 70 and 90 percent.

The natural reservoirs of filoviruses are unknown. Although recent research suggests that they have been lurking on the fringes of human activity for centuries, Marburg and Ebola joined a new category of "emerging viruses" threatening to eclipse more familiar infectious diseases.

A strain of Marburg arrived in the Soviet Union a decade after it was first isolated, during one of our periodic global searches for promising material. It wasn't clear from the records whether we obtained it from the United States or directly from Germany, but it was immediately added to our growing collection of viral warfare agents. We were already investigating a number of microorganisms that weaken blood vessels and cause hemorrhagic fevers, such as Junin from Argentina and Machupo from Bolivia. Marburg quickly proved to have great potential.

Ustinov had been conducting a series of experiments with guinea pigs and rabbits to monitor the effects of increasingly higher concentrations of Marburg. The injection of such a highly concentrated dose directly into his thumb meant that he now had hundreds, perhaps thousands of times more particles of the virus coursing through his body than any of the victims in Germany. I thought his chances of survival were near zero.

I called our biosafety department and asked them to send technicians at once to the viral center of the Ministry of Defense in Zagorsk, where scientists had isolated a Marburg antiserum. Then

I instructed the Ministry of Health to send a team of physicians to Siberia with the serum.

It was a shot in the dark. Koltsovo was four hours away by plane and the next flight from Moscow wasn't until later that night. Even if they made the flight, they would arrive nearly two days after the initial infection—an eternity for Marburg. Zagorsk had only a few hundred milliliters of antiserum on hand.

Kalinin was in a meeting when I asked to see him. Tatyana took one look at me and hurried me into his office. He dismissed his visitors, and I gave him the scanty details I had of what had happened.

"I'm waiting for a cryptogram from Vector, but it looks to me like we have a dead person on our hands," I said.

Kalinin turned pale.

"You don't think he can be saved?" he asked.

"I can't be too optimistic."

"We'll have to tell the higher levels," he said with a grimace.

I couldn't blame him for being as preoccupied with our superiors' reaction as with Ustinov's well-being. We both knew that any major accident would put Biopreparat at risk. Memories of the Sverdlovsk catastrophe's effect on the army's program were still vivid. This was less than two years after Chernobyl; the Soviet Union was in no mood for a new disaster.

Yet the state shared the blame for Ustinov's accident. My visits to Vector had shown me under what pressure we were placing our best scientists. Sandakchiev had never ceased to complain about the inhuman pace at which his workers were being driven. It was dangerous, as well as scientifically unsound. No technician should have worked long hours with such a contagious organism. People tired easily in the heavy protective suits required for Zone Three. Their reflexes slowed down, and it was easy to become careless. Adding to our problems, Marburg research had begun at Vector before a supply of antiserum was on hand.

Ustinov's illness lasted nearly three weeks. Throughout that time, none of his colleagues was allowed to stop working.

———

Sandakchiev's cryptogram arrived early that afternoon. It was long, detailed, and bleak.

Ustinov had been injecting Marburg into guinea pigs with the help of a lab technician, working through a glove box. He was not in a full space suit and was wearing two thin layers of rubber gloves instead of the thick mitts normally required for such work in Zone Three. The gloves provided the flexibility to control the laboratory animals, who will otherwise squirm and try to wriggle out of a technician's grip.

Our rules required that animals targeted for injection be strapped to a wooden board to hold them securely in place. That day, Ustinov wasn't following procedure. He decided to steady the guinea pigs with his gloved hand. Perhaps he thought it would help calm them. Or perhaps he was in too much of a hurry.

The technician became distracted and nudged him accidentally. Ustinov's hand slipped just as he was pressing down on the syringe. The needle went through the guinea pig and punctured his thumb, drawing blood.

The needle went in no farther than half a centimeter, but the faint spot of blood indicated that liquid Marburg had entered his bloodstream. As soon as he realized what had happened, Ustinov called the duty supervisor from the telephone inside the lab.

From then on, the procedures established for such emergencies were followed to the letter. Doctors and nurses dressed in protective suits were waiting for him as he emerged from the disinfectant shower. They rushed him to the small hospital in the Vector compound, a twenty-bed isolation facility sealed off from the outside with thick walls and pressure-locked doors.

Physicians did what they could to make Ustinov feel comfortable while waiting for the antiserum to arrive from Moscow. He was in no doubt of the danger he faced, but there were periods when he believed he could escape alive. He was lucid enough to describe what had happened in precise scientific detail and to calculate the exact amount of Marburg coursing through his veins. His wife hurried over from her lab, but neither she nor their children were permitted inside the hospital. She was later allowed a few private visits, until the sight of her suffering husband became too much to bear.

Every day for the next fourteen days the cryptograms arriving at my office in Moscow described the evolution of Ustinov's disease in dry, clinical language. Attending physicians and colleagues later supplied the human details.

Ustinov at first maintained his sense of humor, joking with nurses and occasionally planning his next experiments aloud. Within a couple of days he was complaining of a severe headache and nausea. Gradually, he became passive and uncommunicative, as his features froze in toxic shock. On the fourth day his eyes turned red and tiny bruises appeared all over his body: capillaries close to his skin had begun to hemorrhage.

Ustinov twitched silently in his bed while the virus multiplied in his system. Too tired to speak, or to turn over, or to eat, he would drift in and out of consciousness, staring for long periods of time at nothing. Occasionally, lucidity would return. He called for paper during those brief moments to record the progress of the virus as it foraged through his body. Sometimes he burst into tears.

On the tenth day, his fever subsided and he stopped retching. As brilliant a scientist as he was, Ustinov began to entertain the delusion that he was improving. He started smiling again and asked about his family.

The cryptograms describing the disease's remission inspired some in our office in Moscow to hope for the impossible. But I was matching the progression of Ustinov's symptoms with clinical reports of the 1967 Marburg outbreak, and nothing in those reports gave me any reason for confidence.

I gave a daily briefing on Ustinov's condition to Kalinin. He passed the information on to senior officials in the Kremlin.

By the fifteenth day, the tiny bruises on Ustinov's body had turned dark blue, and his skin was as thin as parchment. The blood pooling underneath began oozing through. It streamed from his nose, mouth, and genitals. Through a mechanism that is still poorly understood, the virus prevents normal coagulation: the platelets responsible for clotting blood are destroyed. As the virus spreads, the body's internal organs literally begin to melt away.

Shuddering bouts of diarrhea left rivers of black liquid on his sheets. The scraps of paper on which he had been scribbling his symptoms and which the nurses had gingerly carried out to transcribe each day no longer littered the floor. There was nothing more to write. Everything was unfolding before his doctors' eyes.

The filoviruses were already multiplying by the billions inside Ustinov's tissues, sucking out their nutrients in order to clone copies of themselves. Each viral particle, or virion, forms a brick that pushes against the cell walls until they burst. The cells then sprout wavering hair-like antennae that home in on their next target, where the process of foraging and destruction blindly repeats itself.

Ustinov lapsed into long periods of unconsciousness. When he was awake, some say he exhibited uncharacteristic signs of rage. According to some witnesses, he complained about his heavy workload. Others insist this never happened. While it is true that viruses can affect certain characteristics usually associated with personality, it is possible that Ustinov's behavior was magnified to send a message of protest to Moscow. How else could such a message be delivered without fear of retribution?

The doctors from the Ministry of Health arrived early in the first week with the antiserum. To no one's surprise, it proved useless. Antiviral drugs such as ribavirin and interferon were also tried. Hemorrhagic fevers can sometimes be treated with whole-body blood transfusions, but the medical team concluded that it would in this case be ineffective.

A long cryptogram arrived in my office on April 30, describing Ustinov's condition that day. As I read through it, I noticed that the symptoms appeared worse than usual. I sat up in my chair when I reached the final line: "The patient died. Request permission to conduct an autopsy."

Though I had been expecting it, the news came as a shock. I walked into Kalinin's office and told him the ordeal was over.

"They want to conduct an autopsy," I added.

Kalinin was expressionless.

"I'll inform everyone," he said, and turned back to the file he was reading. He didn't ask after Ustinov's widow or his colleagues at Vector. It was time to move on.

I don't know how the senior levels of our bureaucracy reacted to Ustinov's death, but no condolence letter was ever sent to his widow. Sandakchiev asked us for ten thousand rubles as special

compensation for his family in addition to the normal pension survivors were entitled to. It was a princely sum in those days, and
Kalinin balked at first, but he finally approved the request.

Even after death, Ustinov was imprisoned by the virus that had
killed him. The risk of contagion made normal interment impossible, so his corpse was covered with chloramine disinfectant and
wrapped in plastic sheeting. The remains were placed inside a
metal box, welded shut, and fitted into a wooden coffin. Only then
was it safe to lay him in the ground.

The funeral was over quickly. Sandakchiev delivered a brief eu-
logy beside a marble gravestone, which, in the Russian tradition,
bore an engraved image of Ustinov and the dates of his birth and
death. The small group of mourners included Ustinov's immediate
family, his closest colleagues, and a cordon of KGB agents who had
worked frantically to keep the circumstances of his illness secret.
No one came from Moscow.

Regulations prohibited the circulation of any reports about accidents, fatal or otherwise, but news of the tragedy spread quickly
through The System. An investigation by the Ministry of Health
and the KGB concluded that the principal person at fault was the
victim himself, who had not followed proper safety rules.

A flood of administrative decrees began to inundate Biopreparat facilities around the country with urgent safety warnings.
Managers were ordered to upgrade biocontainment facilities and
to report on their progress within ten days, just as they had been
following the accident at Sverdlovsk. Like Sverdlovsk, no connection was made between the warnings and the incident that caused
them.

Ustinov was not the last victim. A pathologist from the Ministry of Health on the team conducting Ustinov's autopsy fell sick
after pricking himself with the syringe he had used to extract bone
marrow. The pathologist, identified in our archives as "V," went
through the same agonies as Ustinov, though it was reported that
he had received a much smaller dose of Marburg. After a month
and a half in the Vector isolating hospital, his condition improved.
When he suddenly took a turn for the worse, he was transported

to Moscow. Biopreparat was never informed of his fate, but I learned through unofficial channels that he died soon afterward.

A virus grown in laboratory conditions is liable to become more virulent when it passes through the live incubator of a human or an animal body. Few were surprised, therefore, when samples of Marburg taken from Ustinov's organs after his autopsy differed slightly from the original strain. Further testing showed that the new variation was much more powerful and stable.

No one needed to debate the next step. Orders went out immediately to replace the old strain with the new, which was called, in a move that the wry Ustinov might have appreciated, "Variant U."

At the end of 1989, a cryptogram from Sandakchiev arrived in my office with the terse announcement that Marburg Variant U had been successfully weaponized. He was asking for permission to test it.

Construction at Vector was running far behind the schedule set out in Gorbachev's last decree, and test chambers were still not ready. There were only three other spots where Marburg could be tested: Omutninsk, Stepnogorsk, and a special bacteriological facility at Obolensk, in the Moscow region. Obolensk had to be ruled out because it was too close to the capital, and Omutninsk was just embarking on tests for a new plague weapon. That left Stepnogorsk.

The facility had never been used to test viral agents before. Colonel Gennady Lepyoshkin, who had replaced me as the director of Stepnogorsk, reminded me of that heatedly when I ordered him to prepare the facilities for a Marburg test run.

"It's just too dangerous," he insisted.

The man who had once joked about Nikolai Chernyshov as the "guy who killed a lot of people" in Sverdlovsk was now a sober-minded manager. I respected his views, but orders were orders.

"Don't argue with me," I said. "It has to be done, so do it."

A brace of bomblets filled with Marburg and secured in metal containers was sent on the long journey by train and truck from Siberia to Kazakhstan, accompanied by scientists and armed guards. It took nearly twenty-seven hours for the shipment to reach Stepnogorsk. Another caravan with twelve monkeys followed shortly afterward.

I went to Stepnogorsk twice to supervise the test preparations. It was less than two years since I'd left there for Moscow, but the facility had expanded so much that it was almost unrecognizable.

After testing the weapon in explosive chambers, we applied it to the monkeys. Every one of the twelve monkeys contracted the virus. They were all dead within three weeks.

In early 1990, Marburg Variant U was ready for approval by the Ministry of Defense.

Our scientist had found it more difficult to cultivate Ebola than Marburg—they were not able to reach the necessary concentration—but by the end of 1990, the long-term problem of cultivation had been solved and we were close to developing a new Ebola weapon. Meanwhile, at Zagorsk (Sergiyev Posad) military scientists were putting the finishing touches on new Lassa fever and monkey pox biological weapons.

SECRETS AND LIES

11

THE INSTITUTE OF ULTRA-PURE BIOPREPARATIONS

Leningrad, 1989

"Nikolai Frolov is on the line," my secretary dashed in to tell me early one Monday morning toward the end of October 1989. "He says you have to talk to him at once!"

I pushed the papers on my desk away and felt more than anything like putting my head down and going to sleep. Since Ustinov's death and the testing of Variant U, we had not been given a moment's rest.

A special conference of institute managers and senior staff was to begin the next day in Protvino, a small town just outside of Moscow. More than one hundred people were due to attend—nearly all of my senior staff. I wasn't looking forward to it. We were behind in almost every project, and I knew none of the managers would enjoy the stern lecture I had prepared for them.

Urgent coded messages had been sent to each director with details of the meeting time and place. I had been deluged with questions all morning. I braced myself for yet another complaint. Frolov was deputy director of the Leningrad Institute of Ultra-Pure Biopreparations, one of our most important research facilities. His boss, Vladimir Pasechnik, was one of our top scientists. I picked up the phone.

"We've got a problem," Frolov said. He sounded strained.

"What problem could there be?" I said, trying hard to put warmth in my voice.

"Pasechnik hasn't come."

"Hasn't come? You mean he's not yet in Protvino? Don't worry, it's okay if he's a little late."

"No, no!" Frolov nearly yelled into the phone. "I mean, he hasn't come back from France."

"France? What is he doing in France?" I almost laughed, thinking this was a strange practical joke.

"But you sent him there. You gave him permission to go."

All at once, I remembered. Six months earlier, during one of my official visits to Leningrad, Pasechnik told me that he had been invited by a French manufacturer of pharmaceutical equipment to visit its facilities in Paris. Their new line of fermenters might be worth our while to investigate, he intimated. I agreed.

"Why not go?" I told him. "It will be nice for you to visit Paris. You've been working hard."

A few months later he phoned to remind me of the trip. I was surprised. I thought he'd already gone.

"I was too busy," he explained quickly. "I just wanted to make sure it was still all right with you for me to go."

It was now October, and I'd heard nothing since. I'd assumed that Pasechnik's vacillating travel plans had finally taken him to Paris and that he had long since returned to Leningrad.

"Can you explain to me what is going on?" I asked Frolov as calmly as I could.

The story emerged in a torrent of excited words. At times, Frolov sounded as if he couldn't quite believe what he was telling me.

Pasechnik had flown to France the week before with a colleague from the Leningrad institute. Their meetings had gone well, and the occasional telephone conversation with the home office suggested that they were enjoying themselves. My cryptogram informing senior staff of the general meeting had arrived in Leningrad a few days earlier, and Frolov had telephoned the details to Pasechnik in Paris.

"The two of them were staying in a nice hotel just outside the city," Frolov said. "They were booked to come back Saturday. But after your message, Pasechnik told his assistant to book an earlier flight Friday because he wanted to prepare for the meeting. He said the assistant was welcome to stay on for the extra day, as they had originally planned.

"So early Friday morning, the guy walked into Pasechnik's room and found him in bed, fully dressed. He looked like he hadn't slept all night. There were cigarette butts all over the floor—and Pasechnik doesn't smoke. The guy was shocked. He said, 'Director, you'd better get ready, you're going to miss your flight,' and Pasechnik got up slowly and mumbled, 'Thank you.' He was like a man in a daze.

"He walked over to his assistant and hugged him and said 'proshchai' [farewell], which was a little surprising instead of the usual do svidaniya [good-bye]. The next day, the assistant took the flight back to Moscow that they'd planned to take together. He found Pasechnik's wife waiting at the airport.

" 'What are you doing here?' he said, and she told him she was waiting for Vladimir. They waited together for the next flight from Paris on Sunday, but Pasechnik wasn't on that one either. That's when I decided I'd better call you."

I listened to the entire story with a knot tightening in my stomach. There were only two possibilities. Pasechnik had been in an accident, or he was alive and not coming back.

I thought back to our last meeting in Leningrad. We had spent a long, tiring day going over various projects. Pasechnik seemed sad and a bit depressed as he drove me to the railway station, where I planned to catch the overnight train back to Moscow. I asked him if anything was wrong. Posing such a personal question to a man like Pasechnik was risky. He was one of our senior scientists, twelve years older than me, and had always been somewhat aloof. I worried he might take offense.

"Kanatjan," he had answered, looking at me sadly, "can I be honest with you?"

"Of course."

"It's like this, I'm fifty-one years old, and I'm going through a strange time in my life. I don't know if I've accomplished what I want to. And they're going to make me retire soon."

It was true: the mandatory age for retirement in all weapons programs was fifty-five. But I clapped him heartily on the shoulder.

"I don't know what you're worrying about," I laughed. "Four years is a long time, and they could be your best years!"

He smiled thinly, we shook hands, and I boarded my train.

I might have picked up the distress signals from that conversation or from his wavering Paris plans if I had not been so preoccupied with work. But at Biopreparat we didn't spend time thinking about staff problems. Nor did we concern ourselves with the insecurities of our top managers. Now I was faced with a crisis that would affect not just the morale of personnel but the entire direction of our scientific program.

The Institute of Ultra-Pure Biopreparations had been one of the crucial links in our network since its establishment in the early 1970s. Under Pasechnik's leadership, it had provided many of our breakthroughs in weapons production. One of its most notable contributions was a milling machine that used a powerful blast of air to turn bacterial and viral mixtures into a fine powder. Nothing like this "jet-stream" machine had ever been built before, at least so far as we knew. It was intended to replace the heavy ball-bearing mills used for decades by the Ministry of Defense and to become a standard fixture at all of our production plants.

Work was also done on new approaches to drying and micro-encapsulation—the process of covering agents in polymer capsules to preserve and protect them from ultraviolet light. Highly pathogenic agents were forbidden inside the city limits, so the focus of the institute was on developing new processes and equipment.

One of Pasechnik's most important projects was the modification of cruise missiles for the delivery of biological agents. The Leningrad scientists were asked to analyze the efficiency of aerosol clouds sprayed from a "fast-flying, low-altitude moving object" containing one or more twenty-liter canisters of liquid or dry agent. They designed a moving platform to release canisters as the missile passed over successive targets. The canisters would break apart on impact with the air.

Cruise missiles have revolutionized warfare. With onboard electronic guidance and mapping systems that enable them to fly close to the ground and thus avoid most radar defenses, they can be launched from the air, land, or sea at great distances from their targets. Harnessing them for our use would dramatically improve the strategic effectiveness of biological warfare. Cruise missiles would require far smaller quantities of biological agents than intercontinental missiles and would do just as much damage. And they would increase our capacity for surprise. Multiwarhead intercontinental missiles can be detected by electronic surveillance minutes after they are launched. Planes can be detected by ground observers, giving civil defense and medical teams time to ascertain that an attack has occurred, determine what kind of agent was used, and mobilize for treatment. A cruise missile would offer little advance warning.

This research continued through my final years at Biopreparat. I do not know what came of it.

If Pasechnik's midlife crisis had driven him to defect, Biopreparat had lost a genuine scientific pioneer—and some of our most delicate secrets were in danger of being exposed to the outside world. In the fifteen years of its existence, not one scientist or technician had ever defected from the agency.

I told Frolov not to say a thing and dialed the extension number for Savva Yermoshin, our KGB chief.

"Savva," I said, "we've got a problem."

"You always have problems," he laughed.

"I think Pasechnik has defected."

There was a dead silence on the other end of the line. Then one word. "Shit."

"We'd better see Kalinin at once," he added, after another long pause.

"That's why I called you. I want you to help me break the news."

Kalinin was talking with Valery Bykov, the minister of medical industry, when we walked in.

I don't remember whether Yermoshin or I spoke first, but I remember the look that passed between Kalinin and Bykov. It was as if they'd just heard about the death of a close relative.

I told them quickly what Frolov had told me.

Kalinin spoke first.

"Who gave him permission to go?"

"I did," I said. "But I told you about it."

Shortly after Pasechnik's initial request, I had informed Kalinin of the invitation from the French company. I had full authority to grant permission for such trips, but Kalinin had told me to keep him abreast of staff movements.

"I don't remember that," Kalinin shot back, glancing at Bykov. "You didn't tell me a thing!"

I felt an involuntary shudder. Kalinin was making it perfectly clear that I was on my own. He looked uncomfortable nonetheless.

Bykov seemed suddenly to enjoy the awkward position in which his rival had been placed. A veteran of Kremlin power struggles, he knew how to take advantage of such situations.

"Who prepared the cryptogram asking Pasechnik to come back to Moscow?" he asked in his deep official voice, as if beginning an investigation.

"I did," I said. "But we sent the same message to all the directors for today's meeting."

"Who signed it?"

"Smirnov," I said, naming one of Kalinin's assistants. I had given him the job of sending the notices out to directors as I'd been too busy.

This was a small mark in my favor. If I had actually signed the advisory, it would have added fuel to the conspiracy theory that seemed to be developing: Alibekov gives Pasechnik permission to go to Paris; then Alibekov signs a telegram telling him to return early. What else could this be but a cleverly coded warning to Pasechnik to stay away? Needless to say, logic wasn't the theory's strong point. But the Soviet mind computed evidence in odd ways.

Bykov was relentless. He asked me to repeat the story again. Then he asked Yermoshin to give his own version, which of course was the same as mine. I held back from describing my impressions of Pasechnik's mood in our previous encounter, since it would have made things look worse. I would then have to explain why I didn't inform anyone earlier of the director's erratic behavior.

Finally, Bykov sat down.

"Mikhail Sergeyevich is going to learn about this," he said, referring to Gorbachev. "I can't keep it from his people in the Kremlin. He'll probably know in a couple of days. You will have to be prepared."

"For what?" I asked.

"Somebody will have to be the scapegoat," Bykov said calmly. "If Gorbachev says we have to punish whoever was responsible, you'll be the one. Of course, if he takes it well, you can live out the rest of your days in happiness."

I nodded dumbly. There was nothing to say.

When I came back from the Protvino conference a few days later, Yermoshin was waiting in my office.

"Have you heard from Pasechnik?" he asked abruptly.

"No, why?"

Yermoshin looked down at his fingernails.

"Well, we think we know where he is."

"How did you find him?"

"We used a psychic," Yermoshin explained. "We showed him a photo of Pasechnik, and he stared at it for a long time until he told us that your man was on some sort of island, a big island, very close to the sea."

"An island?" I said, puzzled.

"Yes of course," Yermoshin went on. "And he said there was a large old building, with two or three men working with him."

I started to smile. I had never dreamed the KGB was interested in extrasensory perception.

"Come on, Savva," I said. "You've got to be kidding. Do we really need this kind of spiritual activity in an organization like ours?"

"Look," he said, suddenly irritated. "This is very serious. The man has done extremely successful work for us in the past."

I dropped the matter because Yermoshin seemed so sensitive, but the curious thing was that the psychic was right.

Pasechnik had gone to England.

In January 1995, long after I had defected, I was invited by the British government to discuss biodefense issues. During a break, several British officers came up to me and we began to talk through an interpreter about Pasechnik, whom I had not seen since 1989. The mood was light, and I casually told them the story of the KGB and its psychic. They didn't laugh.

"But that's exactly where we had him," one of the officers said. "We wanted to keep him secure, so we brought him to an old house on the coast."

The KGB psychic was either remarkably talented or he had remarkably good contacts. Even at the time, I suspected that Yermoshin had been ordered to let me in on the psychic's "secret" to check my reaction. Bykov or Kalinin must have been determined to catch me up. If I had not shown surprise at the news about the island, it would have been proof of guilt. I was angry with Yermoshin for agreeing to be part of such a clumsy trick.

At the end of the week, Kalinin called to say I was "safe." Gorbachev had ordered us to take whatever measures necessary to protect Biopreparat from further damage, but he had not ordered any disciplinary measures against our staff.

A few weeks later, a damage control team met in Kalinin's office. There were two senior KGB officers and several people from our agency, including Yermoshin and Vladimir Davydov, a military engineer in charge of "organizational matters" in all our facilities. Davydov was not one of my favorite people: he was cruel to his subordinates and seemed to me too eager to do what was asked of him.

A consensus was reached quickly. Everything relating to secret defense work at Pasechnik's facility would have to be destroyed. The Institute of Ultra-Pure Biopreparations would become civilian in fact as well as in name. This would leave a major hole in our program, but there was no alternative.

The talk turned to Pasechnik himself, and the atmosphere heated up. He was labeled a traitor, a turncoat, a weakling.

"We have to do something about him," Davydov declared.

We looked at him expectantly.

"There's only one thing to do. He has to be killed."

There was a communal intake of breath, and several people

began to fidget in their seats. Even Yermoshin seemed uncomfortable. Kalinin stared out the window.

I was dismayed. "We can't do that," I said.

I was angry at Pasechnik for having put me in such a difficult situation, but I couldn't tolerate the idea of assassinating him.

One of the KGB colonels spoke up.

"I'd like to stop this discussion," he said softly. "Nobody is going to say anything about killing here."

A chill came over the room. I think we all got the message: if an assassination was required, the KGB needed no advice from amateurs.

I don't know if any attempt was ever made on Pasechnik's life, but he remains very much alive today in the United Kingdom.

Pasechnik's defection foiled our strategy to head off mounting suspicion about our program. We were told that Americans had begun to demand entry to our labs as early as 1986, charging that we were violating the Biological Weapons Convention. Their requests unsettled Moscow. It was difficult to deny access, even though the treaty contained no explicit requirement for visits. And once knowledgeable foreign scientists set foot in one of our installations, our secret would be out. Or so we thought.

In 1988, a year before congressional hearings into biological warfare began in Washington, Gorbachev signed a decree, prepared by the Military-Industrial Commission, ordering the development of mobile production equipment to keep our weapons assembly lines one step ahead of inspectors.

When I became first deputy director in 1988 I was put in charge of inspection preparations, an assignment that soon crowded out my other duties. One of my responsibilities was to act as the agency's representative on a special commission at the Ministry of Foreign Affairs. The so-called Inter-Agency Commission was established to "advise" the foreign minister on arms control, but it was primarily concerned with responding to American complaints about weapons-treaty violations. Every state organization connected to the biological warfare program had a representative on the commission, including the army's Fifteenth Directorate, the

Military-Industrial Commission, the Ministry of Defense, and the Soviet Academy of Sciences.

By 1989, so many charges were flying between Washington and Moscow that the commission was forced to meet almost every month.

The meetings at Smolenskaya, the foreign ministry headquarters in the capital, were chaired by Deputy Foreign Minister Vladimir Petrovsky. Neither he nor anyone else at the foreign ministry was officially told of the existence of our program. Even Foreign Minister Eduard Shevardnadze, a full member of the Politburo and a confidant of Gorbachev's, was kept out of the loop. Although we invariably presented ourselves as "experts in biological defense questions," many of the senior bureaucrats at the foreign ministry seemed to guess what we were up to.

Nikita Smidovich, a sharp young foreign ministry department chief, sometimes sat in for Petrovsky as acting chairman. During one meeting, he brought up the latest American diplomatic dispatch.

"They claim we have a biological warfare facility in the Kirov region, at Omutninsk," he said.

General Valentin Yevstigneyev, commander of the Fifteenth Directorate, looked shocked.

"Absolute nonsense," he said. "The only facilities we have in Kirov are for developing vaccines."

They looked in my direction.

"Well," I said, "we make biopesticides in Omutninsk."

Smidovich grinned at me.

"Come on," he said. "I'm not stupid. You can at least tell us the truth."

"I don't know what you're getting at, Nikita," I insisted. "I am telling the truth."

He shook his head.

"You guys really shouldn't bullshit me," he said.

We all pretended not to understand what he meant. It was obvious that Soviet diplomats couldn't be told that they were being used to stage an elaborate coverup.

A special task force to coordinate our various deception plans was set up at the Moscow Institute of Applied Biochemistry. The institute had no connection with biochemistry: its function was to design and manufacture equipment for our labs.

The task force was given the equivalent of $400,000 to create a cover identity for our operations and to demonstrate our "civilian" character. It drew up blueprints for a fictional biodefense plant, complete with government orders mandating the highest biosafety protection. We wanted to be able to explain why Soviet installations contained ten of thousands of square meters allocated as Zone Three areas, when pharmaceutical operations in other countries rarely operated under such stringent conditions. The United States has only two facilities designed for work at Biosafety Level Four, the equivalent of our Zone Three.

If an observer happened to visit one of our civilian labs that actually produced vaccines and noticed that their protection levels were not as high as our blueprints called for, we planned to tell them that these labs had been built decades ago and that the Soviet Union had increased its safety standards to provide the best possible level of protection for workers. Who could argue with that?

To support this fiction, we created another special unit to "supervise" the construction of these phantom high-safety facilities and had blueprints prepared. We even hired dozens of civilians to serve as engineers.

We were as clever and resourceful as Iraq would be nearly a decade later when confronted with similar international suspicions.

Russia has had long practice in the art of deceiving outsiders, not to mention its own people. The story of the Potemkin village—in which an obsequious prince erected a string of gleaming settlements to prevent Empress Catherine from noticing the poverty of her subjects—is part of Russian folklore. And there is the legend of the hidden city of Kitezh, protected from foreigners by a cloak of invisibility.

Nevertheless, some of us worried that foreign inspectors would see through our schemes. A report sent to me by Colonel Viktor Popov, director of the Institute of Applied Biochemistry, warned that only the most gullible visitor would accept our claim that the giant fermenters and testing chambers in our facilities were used to make pesticides. I rejected his report. "You haven't been given all this money to tell us what can't be done," I told him. Stung by my criticism, he went back to work. It was true, however, that the most suspicious looking equipment would have to be moved to hidden storage facilities.

By 1988, a full year before Pasechnik's defection, we had produced an instruction manual for Biopreparat employees on how to answer queries posed by inspectors. Every conceivable question—What is this room for? Why is this equipment here?—was followed by a prepared reply, which workers were expected to memorize.

I was most concerned about our smallpox project. If foreign inspectors brought the right equipment to the Vector compound in Siberia, they would immediately pick up evidence of smallpox. This constituted a violation of the World Health Organization's resolution, which restricted our stocks of the virus to Moscow's Ivanovsky Institute. We considered transferring those Moscow strains to Vector to establish a plausible reason for keeping smallpox in Siberia, but the Ministry of Health, which controlled the repository, turned us down.

Meanwhile, the Foreign Ministry's Inter-Agency Commission was kept busy responding to an avalanche of American inquiries. Each response, written with the guidance of our "biological defense specialists," was precise, professional, and unequivocal—and each was a lie from top to bottom. The strain of so many lies wore some people down.

At one meeting toward the beginning of 1990, Petrovsky had a broad smile on his face. He told us that he had an important announcement to make. I thought perhaps the threat of American inspections had finally been lifted.

"Our next meeting," he said, "will be conducted by our new deputy minister, Viktor Karpov."

Petrovsky was wearing a bandage on his finger. He picked at it with the absorbed concentration of a small child. For a man who had apparently just been been fired, he seemed unduly pleased with life.

Realizing that we were staring at him in confusion, he looked up with an amused expression.

"I'm free of all this now," he said. "Thank God."

Toward the end of 1989, the American and British ambassadors in Moscow presented a diplomatic demarche to Anatoly Chernyayev, foreign policy adviser to Mikhail Gorbachev. The demarche said the two governments were in possession of "new information" that suggested that the Soviet Union was violating the 1972 Biological Weapons Convention. This could only be Pasechnik.

Ambassadors Jack Matlock of the United States and Roderic Braithwaite of Great Britain were treated to an opaque reply by Gorbachev's adviser.

"There are three possibilities one could assume about the information you are giving me," Chernyayev calmly told the diplomats. "One is that the information is wrong; a second is that Gorbachev knows of this but hasn't told me; and a third is that neither he nor I know."

He promised to "look into the matter."

From that moment, events moved at breakneck speed. Kalinin called me into his office to report that a complaint had been received from Washington and London about our program. I had never seen him so upset.

"We're going to have headaches from now on," he said.

"Shevardnadze is furious. When he found out about this note, they say he went straight to Gorbachev and demanded to know what was going on. Apparently he doesn't like to learn from foreigners about what's going on in his government."

Kalinin shared the army's contempt for the foreign minister, who was then beginning negotiations that would result in the withdrawal of our forces from Eastern Europe. I could imagine Shevardnadze's outrage. A new argument with the West would undercut the steps he and Gorbachev were taking to remake the image of the Soviet Union. Gorbachev had traveled that fall to the Vatican, where he became the first Communist leader to meet the pope. He had offered scant sympathy to the besieged Communist regimes of Eastern Europe in their losing struggles against the forces of democracy.

Kalinin knew, as I did, that Shevardnadze was not part of the small Kremlin circle that had been briefed about our biological weapons program. Only four members of the senior leadership—

Gorbachev, KGB chairman Vladimir Kryuchkov, Defense Minister Dmitry Yazov, and Lev Zaikov (the Politburo member responsible for military industries)—were fully aware of our secret. But it was hard to believe that Shevardnadze didn't harbor the same suspicions about our program as the senior bureaucrats in his ministry.

We took comfort in the fact that there were many things Pasechnik didn't know. He had not been personally involved in weapons production, and much of what he could tell Western intelligence agencies was likely to be hearsay at best, thanks to our internal security regime. Nevertheless, Pasechnik's interrogators would have learned the secret that had been kept hidden for so long: the real function of Biopreparat.

Pasechnik's defection shook the highest levels of the Soviet government. Igor Belousov, deputy chairman of the Council of Ministers and the head of the Military-Industrial Commission, was ordered to prepare a response to the U.S.–British demarche. By February 1990, a draft was ready for the signatures of leading ministers.

Biopreparat wrote the bulk of the document. While declaring that the Soviet Union fully complied with every clause of the Biological Weapons Convention, we conceded that observers might consider some of our activities suspicious. Nevertheless we insisted that all our research into biological warfare agents was conducted for the sole purpose of defending ourselves against potential aggressors. The treaty's ambiguous definition of biodefense work had given us an important loophole.

We also agreed to negotiate a schedule of visits to biological facilities on both sides of the cold war divide. These would not be formal inspections—the treaty didn't require them. But our willingness to open our installations to outsiders would show our sincerity and good faith. None of us really believed that the U.S. government would take this last suggestion seriously. It would force Americans to allow us inside their own bioweapons installations.

The key Soviet leaders, including KGB chairman Vladimir Kryuchkov, Gosplan chairman Yury Maslyukov, Military-Industrial Commission head Igor Belousov, Foreign Minister Eduard Shevardnadze, and Defense Minister Dmitry Yazov, were asked to review and sign the document before it was passed on to Gorbachev for final approval. A formal diplomatic reply would then be sent to the American and British governments.

Kalinin was given the job of obtaining the signatures. He managed to secure everyone's approval, except Shevardnadze's. Inexplicably, the foreign minister dug his heels in. Kalinin was beside himself with anxiety.

One day he came to see me, looking suspiciously relieved.

"They've decided all we need is the signature of Karpov, the deputy minister, since he's head of the disarmament department," he announced. "You can bring the document to him."

I headed through the midday Moscow traffic to Smolenskaya, one of the capital's seven wedding-cake skyscrapers built by Stalin's architects in the late 1940s and early 1950s. I didn't need an armed guard, since there were no state secrets in my briefcase— just a portfolio of lies.

Karpov was absorbed with his papers. He looked up as if surprised to see me, though I knew he had been expecting me.

"What have you brought me, Kanatjan?" he said.

I handed him the paper and waited for him to finish reading.

"You know, young man, I see a troubled future ahead of you," he said at last.

I was taken aback. "What do you mean?" I asked. "Look at all the others who have signed this document. I'm just the courier."

"Kanatjan," he said with a weary shake of his head, "I know who you are and I know what you do. And I know that none of what's written here is true."

"I don't understand what you mean," I said earnestly.

Karpov did not have Nikita Smidovich's sense of irony. He raised his hand to stop me.

"Forget it," he said.

He signed the document and handed it back to me. I hurried back to Samokatnaya Street.

Within a few days a formal diplomatic reply to the Western demarche was delivered to the U.S. and British embassies in Moscow. It was written on Soviet Foreign Ministry stationery and signed by Eduard Shevardnadze.

We were told at Biopreparat that the Americans and British had agreed to keep the Pasechnik affair quiet in return for a "full" Russian response to the diplomatic demarche. Our response was anything but full, but they kept their side of the bargain. Pasechnik's defection became public only after the collapse of the Soviet Union.

Why did our rivals cooperate in guarding our secret? Disclosure of the information Pasechnik gave them would have caused us more harm than a dozen Sverdlovsks. After I settled in the United States, a senior official who had served in President Bush's administration told me American and British leaders believed that a public quarrel would endanger progress in other areas of arms control and perhaps weaken Gorbachev. They were also convinced that their covert pressure would force us out of the biological warfare business.

The West's diplomatic tact may have seemed sensible at the time, but it gave us unexpected breathing space. We continued to research and develop new weapons for two more years.

12

BONFIRE

Obolensk, 1989

Bacteriological warfare is science stood on
its head . . . a gross perversion.
—*from an official paper published by the Soviet Union in 1951*

In a forest clearing on the southern outskirts of Moscow stands a
heavily guarded building. Part of a research complex operated by
Biopreparat at Obolensk, an abandoned village transformed into a
closed city, it housed our "Museum of Cultures." The hundreds of
bacterial strains stored there in small glass flasks provided the raw
material for many of the Soviet Union's groundbreaking experiments with genetically altered biological weapons in the late
1980s.

Building One, a giant glass biocontainment facility with Zone
Two and Three labs, towered above all the other structures at
Obolensk. Five of its eight stories were divided according to pathogens. The second floor was for work on plague. The third was for
tularemia. Higher levels were earmarked for anthrax, glanders, and
melioidosis. Other floors were devoted to work on new industrial
techniques.

In November 1989, a month after Pasechnik's defection, I
joined more than fifty of Biopreparat's senior scientists and military officials in the large and windowless auditorium of Building

One for our annual review of the facility's work. We were not allowed to bring briefcases or bags inside the room. We could take notes, but they were gathered up by security guards after each meeting. We had to get special permission to see them again.

The second-to-last speaker was a young scientist from Obolensk. He approached the lectern to deliver a report on the status of a project known as Bonfire. Few paid attention at first. Work on Bonfire had dragged on for some fifteen years, and most of us had given up hope of ever obtaining results. The project was ambitious. It had been overseen by a brilliant and cantankerous molecular biologist named Igor Domaradsky, who would eventually denounce the entire Soviet biological weapons program. Its goal was to create a new kind of toxin weapon.

Scientists have spent decades trying to manufacture killing agents from the venom of snakes and spiders and the poisonous secretions of plants, fungi, and bacteria. Most nations with biological weapons programs, including the Soviet Union, eventually gave up on harnessing the toxins produced by living organisms. They were considered too difficult to manufacture in the quantities required for modern warfare. In the early 1970s the Soviet government was persuaded to try again, following a remarkable discovery by a group of molecular biologists and immunologists at the Soviet Academy of Sciences.

The scientists had been studying peptides, strings of amino acids which perform various functions in our bodies, from regulating hormones and facilitating digestion to directing our immune system. One important group of peptides, called regulatory peptides, is activated during times of stress or heightened emotion— anger, love, fear—or to fight disease. Some regulatory peptides affect the central nervous system. When present in large quantities, they can alter mood and trigger psychological changes. Some can contribute to more serious adverse reactions such as heart attacks, strokes, or paralysis when overproduced. In a series of trailblazing experiments, the scientists found a way to duplicate in the lab the genes for a handful of regulatory peptides with known toxic properties. One of these was found capable, when present in large quantities, of damaging the myelin sheaths protecting the thousands of nerve fibers that transmit electric signals from the brain and spinal cord to the rest of the body. Unknown in the West, we called it myelin toxin.

As with all peptides, it was hard to obtain enough for useful experimentation. Genetic engineering solved this problem: scientists could synthesize the genes that code for the production of myelin toxin, reproduce them artificially in the lab, and insert them into bacterial cells. If a bacterial strain compatible with myelin toxin could be found, the transplanted genes would multiply along with the bacteria. The project was full of possibility, but the stigma placed on genetic research since the days of Stalin and Lysenko made government support unlikely.

The biologists enlisted the help of Yury Ovchinnikov, who was just then beginning the political crusade that would lead to the founding of Biopreparat. Ovchinnikov immediately recognized the weapons potential of this research. With his colleagues, he drafted a paper calling for the revival of toxin weapons development and sent it to the Central Committee of the Soviet Communist Party.

The paper noted that recent genetic engineering techniques developed in the West made it possible to produce cloned genes as efficiently as bacterial cultures. The apparatchiks couldn't have understood the science, but they were impressed by the caliber of the men who had drafted the proposal. Rem Petrov, a leading immunologist and regulatory peptide expert, now vice president of the Russian Academy of Sciences, was one of its principal authors. The scientists' final argument was irresistible: weapons based on compounds produced in the human body were not prohibited by the Biological Weapons Convention. Funding for Bonfire was quickly approved. Myelin toxin genes created at the Soviet Academy were sent to Obolensk, where research began.

If all went as planned, the Soviet Union would soon have a new weapon, and Russian scientists would at last be able to participate openly in the biotechnological revolution that was sweeping the world.

Genetic engineering arose partly in response to one of the most disheartening developments in modern medicine. Less than twenty years after the discovery of powerful antibiotics, an alarming number of bacteria had discovered a way to outwit them. In an example of nature's own talent for genetic engineering, countless disease-causing microorganisms had spontaneously formed a resistance to the wonder drugs of the 1930s and 1940s.

Antibiotics do not always kill bacteria; sometimes they simply inhibit their growth, allowing the body's disease protection system to overwhelm them. One of the principal differences between our cells and those of bacteria is the presence of a rigid cell wall that protects the bacteria from hostile environments. Most antibacterial agents attack or infiltrate this membrane. Bacitracin, for instance, inhibits the movement of proteins from the cytoplasm to the cell wall, blocking its regeneration. Penicillin and cephalosporins prevent the cell wall from forming, killing the bacteria by leaving it exposed to osmosis. Aminoglycosides, including streptomycin and gentimicin, kill bacteria by binding to their ribosomes and blocking protein synthesis. Erythromycin and tetracyline act in much the same way.

Some antibiotics block or interfere with the bacteria's formation of compounds necessary for growth and reproduction. In the 1930s scientists found that when certain chemical dyes containing sulphur were added to bacterial cultures, the bacteria reproduced at dramatically slower rates. Sulfonamides or sulfa drugs virtually eliminated the threat of pneumonia in Britain after 1935. Subsequent researchers discovered how to inhibit bacterial growth with fungi or molds that could be bred in the laboratory. One of the most effective of these molds was penicillium.

By the 1940s dozens of antibacterial agents were available to physicians for the treatment of diseases ranging from diphtheria to plague, typhus, and tuberculosis. Yet within a few years, some of them began to lose their efficacy as resistant strains of old diseases emerged.

In 1946 the American biologists Joshua Lederberg and Edward Tatum identified one cause of antibiotic resistance and, in the process, created the foundation for the modern science of genetic engineering. Microbes appeared to "learn" resistance to new threats by borrowing genes from one another. When the scientists mixed strains of two microorganisms together, a spontaneous transfer of genetic material occurred. Tatum, Lederberg, and

George Beadle won the Nobel Prize in 1958 for demonstrating that biochemical reactions in microbes were controlled by genes.

Techniques were soon found to manipulate these exchanges. The new processes that were developed changed not only medicine but pharmacology, agriculture, and dozens of other fields. Insulin, for example, a hormone crucial to the treatment of diabetes but produced only in small quantities by the body, could be grown in the laboratory by transferring its genes to bacteria. Human insulin became widely affordable to diabetics for the first time. The genes of corn, rice, and other crops were similarly manipulated to improve the plants' resistance to disease.

News of these developments aroused excitement in the Soviet Union, and envy. Why couldn't our scientists perform just as well? Brezhnev's decision in 1973 to allow genetic experiments under the umbrella of Biopreparat came as an unexpected gift to many Soviet scientists, who had until then been forced to watch the unfolding genetic revolution from the sidelines. The hunger to be on the newest frontier of biology was so powerful that scientists who answered the call to participate in the new program were willing to overlook its connection with weapons-making.

In the winter of 1972, Igor Domaradsky, a molecular biologist and geneticist, was relaxing at a hotel near Moscow when he received an urgent message from the Ministry of Health. He was told that a government car would arrive shortly to take him to an important meeting. Within the hour, Domaradsky found himself in the Kremlin, talking to one of the chiefs of the Military-Industrial Commission.

Domaradsky was offered a job in a mysterious new organization which, he was told, would be devoted to investigating antibiotic-resistant strains of plague and tularemia. As a young scientist, he had made valuable contributions to research on plague. During the 1950s he served as a director of anti-plague institutes in Siberia and southern Russia, where he improved existing vaccines against plague, cholera, and diphtheria. Domaradsky was under no illusion as to the nature of the work he was being asked to do, but he was convinced he could continue his own research under the mantle of the weapons program.

"Our work was directed towards the solution of strictly scientific problems," Domaradsky wrote in a memoir published privately in Moscow in 1995. "It was only later that doubts of a moral nature arose."

Domaradsky, who became deputy director of the scientific advisory council to Biopreparat and represented the organization on the Inter-Agency Scientific and Technical Council, consoled himself at first with the thought that geneticists and biochemists who wanted to remain at the top of their field had no other place to go. "Few of the people who escaped the temptations offered by the government achieved anything in life," he wrote in his memoirs, "or got a chance to work."

The Inter-Agency Council was responsible for coordinating the flow of information between the various branches of government and the state scientific organizations implicated in the Soviet biological weapons program—the Ministries of Health and Agriculture, the Ministry of Defense and of Chemical Industry, the Fifteenth Directorate, and the Soviet Academy of Sciences. It met once every two to three months to discuss the main direction of research and weapons development. The link with the Academy was one of the most important. Four of its institutes were directly involved in biological warfare. Although they didn't develop weapons, they provided Biopreparat in particular with advice based on their fundamental research into pathogenic microorganisms and on their investigation of genetic engineering.

Several of the country's most prominent Academicians were on the Interagency Council—Rem Petrov, an expert in regulatory peptides; Academician Scriabin, an expert on the physiology of microorganisms and an institute director; Academician Mirzabekov, a younger scientist who distinguished himself early for work in molecular biology; and Professor Boronin, who succeeded Scriabin as head of the Institute of Biochemistry and the Physiology of Microorganisms near Moscow.

When I met Domaradsky nearly a decade later, he was a bitter man. An irascible and brilliant theorist with a slight limp due to a childhood bout with polio, he was contemptuous of the organization's military leaders. He had been in the program for so long he could remember when people like Kalinin and Klyucherov, who had briefly served as his deputy, were brash young men. He was convinced they were conspiring to prevent him from pursuing his research.

Few people measured up to Domaradsky's standards, and I was no exception. He sat on the board that reviewed my doctoral thesis and was the only member to criticize my research. But many of us saw more in him to pity than to dislike. He embodied the losing struggle for self-respect waged by many of our most talented scientists locked inside our biological war machine.

Shortly after he joined the Biopreparat advisory council, Domaradsky became involved in plans to establish the genetic institute at Obolensk. He became deputy director of Obolensk in 1973, joining a small group of researchers in the newly built laboratory complex.

An atmosphere of improvisation charged their work. Originally a cluster of red and white brick buildings crisscrossed by dirt roads, Obolensk expanded each month as laboratory equipment was assembled from scientific institutes around Russia. The area around the compound was so sparsely settled that in wintertime elk wandering out of the woods would surprise scientists as they trudged from one snow-covered laboratory to another. The clandestine atmosphere surrounding the project gave it a certain appeal. The scientists saw themselves as pioneers.

Neighbors were told that the Institute of Applied Microbiology was involved in research into infectious diseases, but the high wire fence and heavy gates manned around the clock by troops from the Ministry of Internal Affairs ensured that there would be no casual visitors. Like other institutes, it was identified by a post office box number. Everyone knew that V-8724 meant Obolensk.

Domaradsky recruited a team of scientists from around the country to help him refine the techniques that would be used in Bonfire and Metol, a parallel project centered on the genetic alteration of bacteria to produce antibiotic-resistant strains. He never mentioned Bonfire or Metol in his memoirs, possibly fearing repercussions—both projects are still considered state secrets in Russia—but Obolensk was soon involved in genetic research into the diseases that had been Domaradsky's lifelong preoccupations. Foremost among them was plague.

There are two principal challenges in altering the genetic makeup of disease-producing bacteria. The first is to find the right mechanism for transporting genes into the DNA of another microorganism. The second is to achieve the transfer without reducing the bacteria's virulence.

Domaradsky turned to plasmids to meet the first challenge. Plasmids are strands of genetic material found in bacteria that carry the codes for such things as virulence and antibiotic resistance. They are used in genetic engineering because they can replicate without harming the organisms they come from and can be transferred intact to a new cell.

Domaradsky's scientists found a plasmid with the genes for resistance to tetracycline, one of the most potent and widely effective of all antibiotics. The plasmid was located in a strain of bacteria called *Bacillus thuringiensis,* used to produce biopesticides.

In a petri dish they mixed small quantities of B. *thuringiensis* with anthrax, cultivated the two strains together, and then placed them in a test tube with tetracycline, to see if the anthrax bacteria would survive. The process required endless repetition. It can take months, even years, using such procedures to isolate a strain with the hoped-for resistance. The antibiotic killed most of the anthrax bacilli, but a few cells survived. Most of these had incorporated the antibiotic-resistant genes from *Bacillus thuringiensis* into their own genetic structure. These new cloned cells could now be used to create tetracycline-resistant strains of anthrax and plague.

The second challenge, that of maintaining the virulence of genetically altered material, was more problematic. Despite his talents, Domaradsky couldn't give the Ministry of Defense what it really wanted. The Soviet army wasn't satisfied with weapons resistant to one type of antibiotic. The treatments available for bacterial diseases offered doctors a broad range of choices. The only worthwhile genetically altered weapon, for military strategists, was one that could resist all possible treatments. In 1976, Domaradsky proposed a "triple-resistant" strain of tularemia. He struggled for nearly a decade at Obolensk but couldn't come up with a strain effective across the entire spectrum of antibiotics while retaining its degree of infectiousness.

The program's military chiefs didn't hide their disappointment. Domaradsky offered no apologies for his failure, arguing that science could not be run according to Five-Year Plans. He was reminded that he had first made the grandiose promise to develop a fully antibiotic resistant strain. This elicited his rejoinder that soldiers had no idea how to run a laboratory.

For scientists like Domaradsky, the biological weapons program was both a blessing and a curse. While it provided the money and laboratory space for advanced research, security restrictions ensured that only a small circle of people would ever know the results of their work. Domaradsky patented ten different plasmid transferral techniques and claimed to be the first in the world to isolate the plasmid responsible for the virulence of plague. But his patents and discoveries were locked in classified government archives, where they remain today.

In his memoirs, Domaradsky described the toll Biopreparat's security restrictions took on scientists even in its earliest days. The soft classical music playing from loudspeakers in the labs did little to ease the pressure of constant surveillance. Scientists were forbidden from talking to their families about their work. Their lives were so constricted that they had to spend their vacations together at the same state holiday camp. A lab chief at the Kirov military facility once ordered the windows of his country cottage boarded up so that he wouldn't have to see his colleagues' faces.

The level of paranoia was so high that Biopreparat employees were often barred from attending scientific conferences abroad. Domaradsky found this embarrassing. "I had to think up reasons for rejecting tempting invitations from foreign colleagues," he recalled. "I would have to say I'd broken my leg, or caught something, or had family problems."

On one occasion, he had to get permission to work on a special culture of plague directly from Yury Andropov, then the chairman of the KGB. When he had successfully completed the work, he was asked to bring the results to the Kremlin. Accompanied by an armed guard, he carried a dish with a culture of genetically altered plague through the gates of the ancient fortress like a rare jewel. He solemnly presented the dish to military and Party apparatchiks. It isn't clear what they had hoped to see.

Such absurdities drove him to despair, but his most bitter struggle began in 1982, when Kalinin appointed a new military commander of Obolensk.

Nikolai Nikolayevich Urakov, an autocratic general from the Fifteenth Directorate, had been deputy director of the Kirov facility. He was fond of giving orders in obscure military jargon and had little patience for civilians, especially those he considered malingerers.

Urakov was himself an accomplished scientist. He had received a state award for developing a Q fever weapon and, for as long as I knew him, he never stopped talking about "his" weapon in tender terms. "I wish we could go back to Q fever," he would say nostalgically. "That was a real weapon, but nobody takes it seriously anymore."

Urakov made Domaradsky's life miserable, pressing him constantly about missed deadlines and undermining his authority by bringing young officers in to take over lab work. He even tried to recruit me when I was in Stepnogorsk.

"We could make a great team," he said.

The proximity to Moscow and the chance to work with some of our most creative scientists made this a tempting offer, but I refused. I knew Kalinin wouldn't want me to move away from the weapons production lines.

Meanwhile the tug-of-war inside Obolensk moved to Biopreparat headquarters. I was at Samokatnaya Street one day when the scientist and the general squared off inside Kalinin's office for an argument that could be heard over the entire floor. As I listened outside Kalinin's door, the two men seemed to be on the verge of violence. Domaradsky accused Urakov of "sergeant major" tactics; the general responded in kind. Exasperated, Kalinin finally begged Domaradsky to keep his emotions in check.

"Is this any way for a scientist to behave?"

It was a question that could properly have been addressed to any one of us.

Kalinin eventually chose to back the interests of the military over the prerogatives of science. Domaradsky was no longer at Obolensk when I came to headquarters in 1987. He had been demoted to the position of a lab chief at an institute in Moscow.

It is clear from Domaradsky's memoirs that he believes the military retain control of biological research today. He notes that both Kalinin and Urakov have remained the heads of major scientific institutes and complains that his hopes of pursuing experiments with plasmids have dried up for lack of funds.

Summing up his government career, Domaradsky declares that the genetics program he worked on for so long "justified neither the hopes nor the colossal amount of material investment."

"Essentially nothing remarkable was ever produced," he concludes.

Domaradsky, unfortunately, was wrong. What Domaradsky began, Urakov would finish. He was able to develop multiantibiotic-resistant strains of plague with a far larger spectrum of resistance, sufficient to overcome practically all antibiotic treatments. And another program that Domaradsky had overseen, Project Bonfire, took a surprising turn.

I had been sitting for hours in the Obolensk auditorium when the young scientist stood up to speak. I was too tired to listen with more than passing interest at first as he began to report on his team's latest attempts to transfer toxin genes into various strains of bacteria.

My attention perked when the scientist announced that a suitable bacterial host had been found for myelin toxin. It was *Yersinia pseudotuberculosis,* closely related to *Yersinia pestis.* Lab results had been excellent, and a series of animal experiments had been conducted in secret.

Inside a glass-walled laboratory, half a dozen rabbits were strapped to wooden boards to keep them from squirming free. Each rabbit was fitted with a mask-like mechanical device connected to a ventilation system. This was one of several standard methods of testing aerosols on small animals.

Watching from the other side of the glass, a technician pressed a button, delivering small bursts of the genetically altered pathogen to each animal. When the experiment was over, the animals were returned to their cages for observation. The rabbits all developed high temperatures and symptoms commonly associated with pseudotuberculosis. In one test, several rabbits also displayed signs of another illness. They twitched and then lay still. Their hindquarters had been paralyzed—evidence of myelin toxin.

The test was a success. A single genetically engineered agent had produced symptoms of two different diseases, one of which could not be traced.

The room was absolutely silent. We all recognized the implications of what the scientist had achieved.

A new class of weapons had been found. For the first time, we would be capable of producing weapons based on chemical substances produced naturally by the human body. They could damage the nervous system, alter moods, trigger psychological changes, and even kill. Our heart is regulated by peptides. If present in unusually high doses, these peptides will lead to heart palpitations and, in rare cases, death.

The mood-altering possibilities of regulatory peptides were of particular interest to the KGB—this and the fact that they could not be traced by pathologists. Victims would appear to have died of natural causes. What intelligence service would not be interested in a product capable of killing without a trace?

It was a short step from inserting a gene of myelin toxin into *Yersinia pseudotuberculosis* to inserting it into *Yersinia pestis*, or plague. In the process, we would have a new version of one of mankind's oldest biological weapons.

Traditionally spread by fleas and rodents, *Yersinia pestis* has been responsible for some of the most lethal pandemics in history. For centuries, plague's relentless spread through cities and across countries inspired an awe and horror matched only by influenza and smallpox. One quarter of the population of Europe died of plague in the fourteenth century in an outbreak known as the Black Death. At the height of the Great Plague of 1665 in London, seven thousand people were dying every week. The last major pandemic began in mainland China in 1894 and lasted over a decade, spreading from Hong Kong to port cities around the world. It ravaged Bombay and San Francisco and other cities along the Pacific coast of the United States. More than twenty-six million people were infected. Twelve million died.

The most invasive and virulent disease known to man, plague is one of three infectious diseases subject to quarantine and international regulation. Every case must be reported to the World Health Organization. A single bite from an infected flea can disgorge as many as twenty-four thousand plague cells into the blood or lymphatic system. After a period of incubation lasting between one and eight days, victims will begin to suffer chills and fever while the body rallies its forces to defeat the invaders. The attempt is usually futile. If it is not treated quickly—and diagnosed accurately—the plague bacteria will ravage the body's internal organs, resulting in shock, delirium, organ failure, and death.

Six to eight hours after the first symptoms appear, painful lumps called buboes begin to form under the surface of the skin, increasing in size and darkening as tissues fall prey to infection. Glands swell, causing so much pain, particularly in the neck, groin, and armpits, that even comatose patients have been known to writhe in agony.

The most severe form of the disease is pneumonic plague. Passed from one person to another by as little as a sneeze or cough, the bacteria invade the bronchial system and produce a fatal attack of pneumonia as fluid fills up the lungs, cutting off the supply of oxygen to distant organs. The incubation period for pneumonic plague is short—rarely more than a few days. The symptoms are sudden and often difficult to distinguish from other infectious diseases. An incorrect or late diagnosis can be fatal.

As the plague bacteria are attacked by the body's immune protection system, they release a potent toxin that leads to further collapse of the circulatory system. Death is invariably painful. Victims of pneumonic plague will succumb within eighteen hours of the toxin's release, sometimes going into convulsions and delirium and usually lapsing into a coma toward the end.

In the twentieth century, improvements in urban sanitation and developments in medicine have made outbreaks of plague rare—fewer than two thousand cases are reported on average every year. But the disease continues to surface in rural areas of the western United States—Texas, California, and the Sierra Nevada, where prairie dogs and chipmunks carry the disease. Recent outbreaks have been reported in human populations in India, Africa, South Asia, and southeastern Europe. The disease even struck down U.S. troops in Vietnam.

Since 1948 the most effective treatment against plague has been streptomycin, an antibiotic administered orally or intravenously. Tetracycline, gentamicin, and doxycycline have also been used successfully. The first plague vaccine was developed by a Russian physician, Waldemar M. W. Haffkine, in 1897, during the Hong Kong pandemic. Several improved vaccines have been developed since then, but they are effective only against bubonic plague. Boosters must be taken every six months. Degrees of immunization vary from person to person, and adverse reactions increase with the frequency of vaccinations.

The earliest recorded use of *Y. pestis* in war was in the fourteenth century, when a Tatar army conquered Kaffa, in present-day Crimea, by catapulting the bodies of plague victims over the walls of the town. During World War II, leaders of the Japanese bacteriological warfare program turned to plague because an attack could be concealed as a natural outbreak. But there were drawbacks: when they tried to drop bombs filled with plague from aircraft, the explosion killed the bacteria. The commanders finally settled on a more effective method of delivery: they blanketed the target area with billions of plague-infected fleas.

Americans tried to develop a plague weapon but found that its virulence deteriorated quickly. The bacteria lost virulence so rapidly—sometimes in less than thirty minutes—that aerosols were useless. U.S. bioweaponeers eventually lost interest, but we persevered. Plague can be grown easily in a wide range of temperatures and media, and we eventually developed a plague weapon capable of surviving in an aerosol while maintaining its killing capacity. In the city of Kirov, we maintained a quota of twenty tons of plague in our arsenals every year.

The success of the Bonfire project raised our plague work to a new level. Within the next few months, scientists at Obolensk successfully transferred the gene for myelin toxin to *Yersinia pestis*. A toxin-plague weapon was never produced before the Soviet Union collapsed, but the success of this experiment set the stage for further research on bacteria-toxin combinations. Soon, scientists were studying the feasibility of inserting the genes for botulinum, the most lethal naturally occurring toxin, into bacteria.

In other circumstances, the discovery by Russian scientists that human regulatory peptides could be reproduced in the lab might have been shared widely, even welcomed as a contribution to our understanding of neurological disease. Instead, it was classified top secret and concealed from the world.

The final speaker at the conference was Urakov. As he approached the microphone to deliver his closing remarks, he could barely contain his pride.

"We have overachieved, as usual," he said.

No one could argue with him. Obolensk by then covered so much ground that workers had to take a bus from one section to another. At the time of the conference, it housed about four thousand scientists and technicians. The facility's annual budget of nearly $10 million paid for the purchase of expensive Western equipment—electron microscopes, chromatography devices, high-grade centrifuges, laser analysis machines.

The myelin toxin report was the last in a series of successes reported that day. Another team had developed a genetically altered strain of anthrax resistant to five antibiotics. And there was a new drug-resistant strain of glanders.

Yet Urakov still wasn't satisfied.

"We haven't been looking hard enough at new drugs being developed in the U.S., Great Britain, and Germany," he said. "Remember, our work for the Motherland is never finished."

13

THE FIRST MAIN DIRECTORATE

Moscow, 1990

Samokatnaya Street felt at times like a cloister. Our secrets cut us off from political life in the capital, and we could not take the risk of making close friends outside the program. In our isolation, we forged relationships among ourselves. We visited one another's homes, gossiped about office politics, exchanged stories about our wives and children, and complained about Kalinin.

One man never joined our circle. His name was Valery Butuzov. A tall, gangly fellow in his early forties with a short military haircut, Butuzov gave no one any reason to dislike him. He always had a cheerful greeting ready when you met him in the hallway and smiled easily. Yet he seemed to retreat from closer contact. Butuzov held a Ph.D. in pharmacology. In our organizational charts he was listed as an engineer, but no one understood what he did. Sometimes he disappeared for days at a time.

General Anatoly Vorobyov, Kalinin's deputy in 1987, complained about him all the time.

"The guy doesn't do anything," Vorobyov once grumbled. "I've never seen anyone so lazy."

I was reviewing with him a list of assignments for new personnel, which required approval from the Central Committee.

"Why don't you fire him?" I asked. "We've got plenty of people to fill his place."

The general was silent for a few moments.

"I can't," he said.

"Why not?"

Vorobyov began to shuffle papers on his desk. He looked annoyed.

"That's really no business of yours, Kanatjan," he said. "Don't you have work to do?"

I took the hint and didn't raise the subject again. But I wondered why Vorobyov, the second most powerful manager in our organization, couldn't fire this man.

When I replaced Vorobyov as first deputy chief, I discovered who Valery Butuzov was. He was not an engineer but a colonel in the First Main Directorate, the foreign intelligence unit of the KGB. His Biopreparat position was a cover for activities too secret even for senior management to know. Yermoshin, our KGB head, knew Butuzov's real identity, but he couldn't tell me his function.

"I have no authority over those guys from the First Directorate," he shrugged. "I'm not even supposed to know he's here. You figure it out—the guy's a pharmacological genius."

Butuzov wasn't much older than I was. I started to engage him in conversation whenever we met. At first he tolerated my attentions—he couldn't exactly be impolite to Kalinin's new deputy—but over time, we discovered we had interests in common. We could discuss the latest books and movies and that great ice-breaking subject for men: sports.

He skillfully deflected questions about his work. Still, he was more open about his background with me than he had ever been with Vorobyov. He once told me he had worked as a younger man in the Ministry of Health, within a facility he called the Institute of Pharmacology, in some sort of intelligence capacity.

After one of his prolonged absences, I asked him where he'd been. He looked drawn, as if he hadn't slept for days.

"They wanted me at the lab at Yasenovo again," he said, shaking his head. "Those guys are idiots sometimes."

My interest was piqued. Yasenovo was the KGB's ostentatious modern spy palace, built in a forested enclave on the outskirts of Moscow to house the First Directorate. Yermoshin spoke of it with envy. The rest of the KGB apparatus, including his own Second Directorate (for counterintelligence and internal security), was confined to the gray-walled Lubyanka building in central Moscow. Yasenovo, which some said was modeled on CIA headquarters in Langley, Virginia, had been the private kingdom of Vladimir Kryuchkov, who spent fourteen years there as foreign intelligence chief before becoming KGB chairman in 1988. Its cafeteria served black caviar and smoked salmon, and senior officers could forget the daily strains of running the world's largest espionage agency in an elaborate sports complex and swimming pool. A monument to the "unknown intelligence officer" stood in its central courtyard. But I had never heard of a pharmacological lab at Yasenovo.

In 1989, Kalinin and I went together to a meeting at a covert division of the Soviet Ministry of Health. This division, known as the Third Directorate, was located far from the ministry's downtown headquarters in a pink office building on Leningradsky Prospekt, in northern Moscow. Its director, a scientist named Sergeyev, held the rank of deputy health minister but seemed to have no contact with his superiors at the ministry. We met with him frequently, but I could never understand why. Most questions relating to vaccines and immunization were dealt with by other departments.

That day, we discussed Ustinov's death in Siberia. Sergeyev ponderously and meticulously analyzed the health ministry's involvement in the incident. He went over the reasons for the shortage of Marburg antiserum and the problems involved in shipping it to Vector, even though his own directorate had played almost no role.

As Kalinin and I waited outside for our driver, I vented my frustrations.

"Yury Tikhonovich, why do we always waste our time at this place?" I said. "We are responsible for biosafety at our installations. There's no reason for Sergeyev to get involved, so far as I can see."

Kalinin glanced at his watch. He hated to be kept waiting, especially when it forced him into idle conversations.

"You're half right, Kanatjan," he replied testily. "We don't really need their help on safety, but they are occupied with other things that make it worthwhile for us to keep our association with them."

"What things?" I asked.

He hesitated. He loved to dramatize moments like these.

"If I tell you, you can never mention it to anyone else," he said solemnly.

"Of course," I said.

"This directorate is responsible for a program called Flute," he said, using the Russian word *fleyta*. "Many institutes come under its control."

"Flute?"

He nodded portentously. It was a code name I'd never heard before.

I pressed him further. "Which institutes?"

He mentioned a few. One was the Severin Institute, which he said was located inside an asylum for the mentally ill in Moscow. Another was a pharmacology institute whose full name he wouldn't divulge. It sounded like Butuzov's old institute.

"What is this program for?" I asked.

Kalinin made a slicing motion across his neck.

"Sometimes people disappear," he said.

"What are you trying to tell me, Yury Tikhonovich?"

He looked disgusted with my stupidity.

"I've said enough," he said.

At that moment our car appeared, ending our conversation. I knew it would be dangerous to ask questions in the office, but I began to watch for clues dropped in conversations and paid close attention in subsequent meetings with the Third Directorate. I was curious about Flute.

The Severin Institute, I eventually discovered, developed psychotropic agents to induce altered mood and behavior in humans. Scientists worked with a number of biochemical substances including regulatory peptides, establishing a shadowy link with our

Bonfire program. Another institute controlled by the Third Directorate, Medstatistika, gathered statistics related to biological research around the world. The pharmacology institute specialized
in developing toxins to induce paralysis or death. All were connected in some manner to the Flute program, whose principal aim
was to develop psychotropic and neurotropic biological agents for
use by the KGB in special operations—including the "wet work"
of political assassinations.

Perhaps Kalinin was right. There were things I was better off
not knowing.

Biopreparat had no formal connection with Flute—our mission
was to produce weapons for war—yet we couldn't completely escape it. The techniques we developed for cultivating, isolating, and
cloning the agents in our labs were useful to many other government programs. It became clear to me that Biopreparat, vast as it
was, was part of a larger zone of clandestine scientific research.

If Butuzov no longer worked at the pharmacology institute,
what was he doing at the Yasenovo laboratory?

His office was on the second floor, a few doors down from
mine. Our friendship grew steadily stronger. I can say without embarrassment that I grew to like him immensely.

As we shared more secrets, Butuzov and I became inseparable. We
went fishing on the Ucha River near Moscow, and our families
spent weekends together at my state dacha outside the city. He was
a wonderful cook and a great handyman. He repaired my Zhiguli
two months after I had proudly accepted its delivery from the state
car plant.

"We can't even make cars right anymore," he would say with a
laugh. "I think they leave parts out just to test us."

I visited the modest Moscow apartment he shared with his wife,
daughter, and elderly mother. I could not reconcile the open-
hearted man I knew with the work he did. Over the course of many
conversations outside the office, he told me more about his work.

Butuzov had been transferred from the pharmacology institute
to the Yasenovo lab many years before he came to Biopreparat.
Known as Laboratory 12, it was established in the 1920s by Genrikh Yagoda, a pharmacist who went on to become one of the cru-

elest of Stalin's secret police chiefs. Laboratory 12 specialized in substances that could kill quickly, quietly, and efficiently.

Butuzov was guarded about the lab's work, but he mentioned a few of its "achievements." In the late 1940s, a powdered version of plague was manufactured for use in a tiny toiletry container, like talcum powder. An assassin could approach a target from behind, spray the lethal powder, and vanish before his victim knew there had been an attack. The assassin would of course have to be vaccinated against plague beforehand to protect him from stray particles.

This device was to be used against Marshal Tito, the Communist partisan who became head of postwar Yugoslavia.

Tito provoked Stalin's anger in 1948 with his plan for a Balkan federation that would dramatically reduce Moscow's control over the region. At the last moment, Stalin decided against assassination. Tito lived to take Yugoslavia down the road of nonalignment and died an old man in 1980.

"Why did Stalin change his mind?" I asked.

Butuzov laughed.

"The only person who knows that is Stalin," he said.

Laboratory 12 was kept busy during the 1970s. In September 1978, Georgy Markov, a Bulgarian dissident, was taken to a hospital in London suffering from a mysterious ailment. Before he died, he casually mentioned that a stranger had grazed him with the tip of an umbrella while walking across Waterloo Bridge. Puzzled doctors were unable to trace the cause of death until a Bulgarian émigré in Paris reported falling sick after a similar scrape with an umbrella. When a second autopsy was performed on Markov, the coroner found the remains of a tiny pellet with traces of ricin, a toxin made from castor beans.

The ricin came from Laboratory 12.

Nearly eight months earlier, the Soviet Union had been asked by the Bulgarian government of Todor Zhivkov to help assassinate Markov. Bulgaria's intelligence service passed the request to its Russian counterpart, but the KGB chairman, Yury Andropov, balked at sending his own hit men to do the job. Instead, he authorized a special consignment of ricin to be sent from Laboratory 12 to Sofia. KGB technicians were sent along to train Bulgarian agents. There were several unlucky rehearsals: at least two failed attempts on Bulgarian exiles, including the one in Paris, were made around the same period.

Butuzov eventually told me why he was based at Biopreparat.

"The pharmacology institute worked exclusively with chemicals," he said, "but we decided the biological area was more promising. So they sent me to your shop."

I don't know what Butuzov really thought about his job, but I noticed that as *perestroika* and "new thinking" came to penetrate more of our political life, he seemed less busy. He looked more relaxed than usual, but I think he was also bored.

In the spring of 1990, Butuzov walked into my office and sank into the big armchair across from my desk. He stared for a while at the portraits of scientists hanging on the wall.

"I need your advice on something, Kan," he said casually.

"Sure," I said. "Professional or personal?"

"Professional."

I waited until he spoke again.

"I'm looking for something that will work with a gadget I've designed. It's a small battery, the kind you use for watches, connected to a vibrating plate and an electric element."

"Go on," I said. He spoke in the same casual tone in which we discussed a soccer match. I was fascinated.

"Well, when you charge this element up, the plate will start vibrating at a high frequency, right?"

"Right."

"So, if you had a speck of dried powder on that plate, it will start to form an aerosol when it vibrates."

He looked at me for encouragement, and I motioned for him to continue.

"Let's say we put this assembly into a tiny box, maybe an empty pack of Marlboro cigarettes, and then find a way to put the pack under someone's desk, or in his trash basket. If we were then to set it in motion, the aerosol would do the job right away, wouldn't it?"

"It depends on the agent," I said.

"Well, that's what I wanted to ask you about. What's the best agent to use in such a situation if the objective is death?"

I'm not sure why I went along with him, but I did.

"You could use minimal amounts of tularemia," I said, "but it wouldn't necessarily kill."

"I know," said Butuzov. "We were thinking of something like Ebola."

"That would work. But you'd have a high probability of killing not just this person, but everyone around him."

"That wouldn't matter."

"Valera," I said. "Can I ask you something?"

"Of course."

"Is this a theoretical discussion, or do you have someone in mind?"

A grin crossed his face.

"No one in particular," he said. "Well, maybe there is one person—Gamsakhurdia, for example."

Most people in Moscow knew by then the name of Zviad Gamsakhurdia, the newly elected president of Georgia. Like most of the Soviet republics, Georgia was moving inexorably toward independence. The flamboyantly mustachioed Gamsakhurdia had been a thorn in Moscow's side for years. The son of a prominent writer, he led the republic's human rights movement and publicly accused Moscow of plotting his assassination. Gamsakhurdia was particularly despised in military circles for the campaign he'd led against the Soviet army after a demonstration in Tbilisi, Georgia's capital, that left nineteen people dead in 1989.

Once in power, his humanitarian impulses were eclipsed by his extreme nationalism. Many felt he had become mentally unstable.

He was unpopular in Russia and I didn't like him. I said nothing more, and we moved on to other topics.

Things were busy for several months after that, and I saw Butuzov infrequently. Then, one Sunday, I invited him and his family over to my dacha for a barbecue. While the *shashlik* was grilling and the children were playing, I whispered a question.

"Valera, what happened to that idea of yours, you know, the one about the watch battery and Gamsakhurdia?"

He smiled.

"Oh, that," he said. "Well, to tell you the truth, it never really got anywhere. We had a plan prepared but the bosses finally turned it down. They said it wasn't the right time."

In early 1992 Gamsakhurdia was ousted from office by his former allies, and former foreign minister Eduard Shevardnadze became the president of independent Georgia. A year later, on December 31, 1993, the fiery ex-dissident died in mysterious circumstances in the course of a violent attempt to return to power. His death was reported as a suicide, but some claimed that he had been murdered by Moscow agents, or by one of his political rivals in Georgia.

One of the principal advantages of biological agents is that they are almost impossible to detect, which complicates the task of tracing the author of a biological attack. This makes them as suitable for terrorism and crime as for strategic warfare.

Many former KGB intelligence agents have been hired by the Russian mafiya. Some run their own criminal organizations. They would have ready access to their former colleagues and to the techniques and substances we developed in the Soviet era. The "achievements" of the Flute program would command a good price on Russia's private market.

On August 3, 1995, Ivan Kivelidi, chairman of the Russian Business Roundtable, was rushed to a Moscow hospital from his office, where he was suddenly taken ill. His secretary, Zara Ismailova, was brought to the emergency room a few hours later with a similar unexplained illness. The secretary died that night, and Kivelidi the next day.

Kivelidi was an outspoken critic of several high-ranking officials in the Yeltsin government, whom he accused of corrupt dealings. The Business Roundtable was composed of leading bankers and entrepreneurs who had banded together to put an end to mafiya control of the burgeoning private sector. Of the original nine members, only Kivelidi was left. The others had all been murdered in mob-style shootings, joining a list of more than five hundred victims of contract killings in 1995.

Kivelidi had taken extra precautions at his office and at home.

Earlier that summer, he announced his intention to start a new political party dedicated to cleaning up Russian capitalism.

Detectives at the murder scene reported that they had discovered an unknown substance on Kivelidi's office telephone. They
identified it as cadmium. The deaths of the businessman and his
secretary were then reported as "radiation poisoning," but when I
read news reports of the incident, they reminded me of a conversation I'd had several years earlier with Butuzov about the killing
efficiency of various aerosols.

"We've come up with an interesting new approach," he told me
with some excitement. "Let's say we spray something on the steer-
ing wheel of a car."

"What would you spray?" I asked.

"That's not important for the moment," he replied. "The point
is, the driver would either pick the agent up by inhaling or through
his skin. It couldn't fail."

"It would have to be very stable to keep its virulence," I said.
"You don't know how much time would pass between the moment
you sprayed the agent and the victim's actual exposure."

"We've got it all figured out," he said confidently. "It would
look like a heart attack."

I expressed admiration.

"Oh," he waved his hand casually. "We've developed lots of
better stuff."

Assassination, thankfully, was not part of Biopreparat's mandate,
but Butuzov's presence showed that the KGB continued its close
association with biological weapons research. I was doubly surprised to discover it had decided to play the role of a dove in the
internal debate over our future following Pasechnik's defection.

Yermoshin appeared in my office one day with the stunning news
that KGB chairman Vladimir Kryuchkov had sent a memo to Gorbachev recommending the liquidation of our biological weapons
production lines.

According to Yermoshin, the memo argued that the Pasechnik
affair had put the Soviet Union in an embarrassing and vulnerable
position. The biological warfare program was no longer a secret.
In our diplomatic reply to the U.S. and British governments, we

had been forced to accept the idea of opening up our facilities. Kryuchkov insisted that there was nothing to do but cut our losses. Such a move might even restore our strategic advantage, since it would force the Americans to open up their biological warfare facilities.

This was the kind of shrewd tactic Kryuchkov excelled at. I was also certain no one in the army or the leadership of Biopreparat would go along with it.

"Not everyone in the KGB supports it either," Yermoshin said. "Bobkov for instance." He was referring to the KGB's first deputy director.

"But you don't know Kryuchkov. Gorbachev trusts him completely."

The odd thing was that I found myself agreeing with the KGB chairman.

Like everyone else, I was furious with Pasechnik and believed he had put our security at risk. But where others desperately wanted to preserve the status quo, I saw no choice but to change course. If the Americans and British came to Russia and observed the size of our production lines, we would be forced to abandon them and to dismantle our entire program. Pasechnik knew a lot, but not everything. He was familiar with our research work but knew very little about our production techniques. Why not try to hedge our bets? If we were to dismantle a significant portion of our production facilities, maybe we could preserve our research programs. If circumstances required, we could always recover our strength. So long as we had the strains in our vaults, we were only three to four months away from full capacity.

Yermoshin told me he had been authorized to discuss the KGB memo with officials at Biopreparat and in the military. Kryuchkov was politically savvy enough to realize he needed military support.

Yermoshin wasn't surprised when I told him I thought Kalinin would oppose the plan at all costs.

"That's why I didn't go to Kalinin to begin with," he said. "You're second in command. I thought we could go together to see Bykov."

This was dangerous. Kalinin would regard any attempt to go

behind his back as insubordination. To approach his worst rival would be treachery. But I saw no alternative.

A few days later, Yermoshin and I went to Bykov's office at the Ministry of Medical Industry in central Moscow. We didn't call ahead for an appointment.

Bykov didn't seem particularly fazed by our unexpected visit. He was more intent, as he sauntered into the reception area, on smoothing the wrinkles from his dark blue suit.

"What brings you here," he said, without an ounce of curiosity.

"Valery Alekseyevich, something important has come up that we need to discuss with you," said Yermoshin.

He sighed, glanced at his watch, and waved us into his office.

We remained standing in front of his desk. He did not invite us to sit down.

"Well, get to the point!"

"The point," Yermoshin said, squaring his shoulders, "is that the KGB believes our biological production lines must be cut. I've been authorized by my superiors to seek your support."

Bykov turned to me.

"What do you think?"

"I agree," I replied at once. "The program wouldn't suffer. We can always—"

Bykov cut me off with a wave of his hand.

"It will never happen," he said brusquely, turning back to his desk.

"You can go now." He didn't look up as we left.

14

INSIDE THE KREMLIN

Moscow, 1990

The headquarters of the Military-Industrial Commission, one of the Soviet Union's most powerful institutions, is a nondescript gray and yellow building backed, as if for protection, against the south wall of the Kremlin. I had been going there as much as four times a month since 1988. By March 1990, when I was summoned to a special meeting, it was almost as familiar to me as my office.

Alexei Arzhakov, the slender deputy chairman of the commission, nodded when he saw me walk in. His boss, Deputy Prime Minister Igor Belousov, was sitting at the head of a large oak conference table. As chief of the agency that supervised the nation's military production, Belousov effectively controlled more than two thirds of our industrial enterprises. In the Soviet system, the manufacture of arms and defense-related products was closely integrated with civilian work.

Inside the conference room were some of the most important members of our warfare establishment, including General Valentin

Yevstigneyev, the new commander of the Fifteenth Directorate, and Oleg Ignatiev, head of the commission's biological weapons directorate.

The meeting began on a now-familiar note: in the aftermath of Pasechnik's defection, how should we respond to the American and British accusations?

I kept silent. No one had any new idea to offer, and Kryuchkov's memo wasn't mentioned. The meeting was coming to an ineffectual end when Arzhakov leaned over and asked me to stay behind. I was concerned, but his expression was friendly.

"There are a couple of people here I want you to meet," he whispered, nodding toward two men in the back of the room.

Belousov gave me a meaningful glance as he walked out the door.

The two men pulled up to the table and started taking papers out of identical black portfolios. Their gray suits and demeanor identified them immediately as intelligence officers.

Arzhakov began the session.

"You've heard what we've been discussing," he told them. "We have a serious problem with the United States related to our biological program, and we haven't been able to respond properly.

"I've invited Colonel Kanatjan Alibekov to tell you what he needs from your agencies."

Both men were generals. One was a high-ranking official in the KGB's First Directorate; the other was one of the deputy directors of the GRU, the Red Army's intelligence wing. I was impressed. It was the first time I'd seen representatives of our two principal spy agencies together in the same room.

I had come a long way from the days when my career was at the mercy of our security organs, and I no longer had anything to fear from such men. This thought gave me a curious feeling of satisfaction.

"What we need should be simple enough to get," I said. "The only way we can deal effectively with the United States is by knowing everything it is possible to know about their biological weapons program. The trouble is, there are a lot of gaps in our knowledge."

"Well, there's Fort Detrick," one of them volunteered, referring to the United States Army Medical Research Institute of Infectious Diseases (USAMRIID) in Maryland, where the United States began its weapons research program in 1943.

I cut him off briskly.

"Everyone who pretends to know something about the American program mentions Fort Detrick. That's old news. Do you have anything better?"

The GRU man looked annoyed. "Why don't you tell us what you need?"

"Fine," I said.

I'd begun to suspect that America presented less of a biological warfare threat than our internal propaganda suggested. While I didn't believe they had ended their program, as they announced in 1969, I wondered why they were so determined to get access to our facilities. Surely they knew we would demand the same from them in return. This prospect didn't seem to bother them, which suggested that their program must be smaller than ours. To my mind, this presented another argument for cutting back.

As I spoke, the generals wrote furiously in their notepads.

"First we need the locations and names of all new facilities created in the past twenty years," I said. "We'll need the names of the commanders and their leadership structure. You'll have to find out which biological agents they are working with and what types of delivery systems they've developed. And we need a record of all of their testing."

I wondered if they would tell me that my demands were naïve. Their faces were grave and gave away nothing.

"Give us a couple of weeks," one said.

A few weeks later I was called back to the Kremlin. This time, Arzhakov and the others were absent. It was just the three of us, sitting at the same huge table. The KGB man spoke first, with evident pride.

"Have you ever heard of Plum Island?" he asked.

My spirits fell. "Of course," I said.

Certain U.S. installations had figured in our intelligence reports for years. As deputy chief, I had seen many of those reports.

Plum Island, in New York's Long Island Sound, was used during the war to test biological agents. It had subsequently been turned into a U.S. Department of Agriculture quarantine center for imported animals and food products.

"We also found something in Illinois," said the GRU man.

"I know about that one as well," I said before he could go any further. "It was abandoned as a weapons production site in the 1950s because they couldn't build suitable biosafety conditions. It's being used by a large pharmaceutical company."

The intelligence officers looked dismayed.

"Don't you have anything else?" I said.

They started to mention a few other places, all of which had been discounted as inactive. In exasperation, I cut them off.

"It's obvious you've just gone through your old records," I said. "Plenty of information about these facilities is available in open literature. I don't need intelligence experts to tell me about them."

I excused myself and went downstairs for a cigarette. I paced back and forth in the cold by the Kremlin wall. For a fleeting moment, I wondered if the generals had been ordered to keep material from me. But the KGB chairman himself wanted to shut down our program. His agency would be as determined as I was to find evidence of American activity.

When I returned to the room, the officers had closed their files. We agreed that there was nothing more to say, and I coolly thanked them for their trouble. Inside, I was shaken.

It was impossible to believe that our most important military rival wasn't pursuing an active biological warfare program.

As part of my duties at Biopreparat I reviewed our budget regularly with Gosplan, the state economic planning agency. Every time I visited the block-long building on Gorky Street, the resources available to us seemed to increase. General Roman Volkov, the balding, scholarly official in charge of funding Ministry of Defense programs, practically begged me to look for ways to spend money.

"I've got three hundred million rubles for you in this year's budget," he told me in 1990. "You still haven't supplied me with programs on which to spend them."

When I suggested civilian medical projects, he brushed me off irritably.

"If you give me more suggestions like that, you'll never get any money," he said.

This made little sense to me. Our health-care system was getting worse every day, and conditions in our hospitals were abysmal. The previous year, Biopreparat had shipped boxes of disposable plastic syringes to medical facilities around the country in response to an AIDS scandal in Elista, a small city on the northern steppes of the Caspian Sea. Two hundred and fifty children at the city's main pediatric hospital had been diagnosed with HIV after having been infected by contaminated syringes. Nurses complained that shortages of equipment and staff prevented them from employing adequate sterilization methods. Stories like these were rife throughout the country.

In February 1990, Valery Ganzenko, the head of the medical directorate, came to my office with a bagful of dirty vaccine vials.

"These are being produced at our facility in Georgia," he complained. "Hospitals in the area are sending them back to us because they're not sterile. When I ask the Georgians what's going on they can't explain it, but we just sent them a big grant to upgrade their equipment."

I was responsible for the civilian institutes operating within Biopreparat as well as for our military research program. The supervision of vaccine production and antibiotic development for our state health system occupied almost 50 percent of our official functions. Kalinin paid almost no attention to this part of our agency, and neither did others in the senior staff. We gave civilian managers free rein, and inevitably much of our equipment ended up on the black market. No one seemed to care. But I found myself increasingly drawn to our medical programs and allocated time to them whenever I could.

"Maybe we should take a special trip down to Tbilisi," I suggested to Ganzenko.

He was pleasantly surprised.

"I didn't think anyone around here wanted to spend time on this kind of thing," he said.

We were met at the Tbilisi airport by the director of the facility, who drove up in a black Volga. A pompous man with a thick Georgian mustache, he was determined to treat Ganzenko and me as VIPs. Before I could protest, he had swept us off on a grand tour of the capital, a city of steep streets and elegant wrought-iron balconies.

"Why don't we go to your lab?" I said.

"Later," he said. "You must first enjoy Georgian hospitality."

On our first night he took us to a restaurant where he'd reserved a private room. The table was laden with meat, cheese, fish, and bottles of wine—all in short supply at ordinary food stores in Moscow.

The poverty of the lab we visited the following morning seemed grimmer by comparison. Some of the equipment was over forty years old. Workers used tiny ovens to culture vaccines. Our host was unapologetic, insisting that Biopreparat's funds were being spent on wages and operating costs.

I knew he was lying as soon as I began talking to the staff. Most of the three hundred employees were women. They told me they were earning such pitiful salaries that they couldn't even afford lunch.

At a general meeting later that day, I announced that the lab would have to be closed.

"The medicine you produce is not acceptable," I said. "It cannot be used to treat people. We have made plans to reassign your production quotas to our laboratories in Ufa and Leningrad."

Everyone began to shout at once. Some of the women sobbed. In broken Russian they complained there was no other place to find work: their husbands were gone, they had hungry children. I was stunned by their despair. I had not witnessed this level of destitution since leaving Kazakhstan.

"I'll give you another chance," I finally conceded. "We'll keep the lab open to see if things improve, but one thing will have to change now."

I pulled out a sheet of paper and began writing.

"With this paper," I said, "I'm firing your director and replacing him with his deputy."

All of the director's charm rapidly vanished. He accused me of stealing his people's patrimony and vowed he would appeal to the government of Georgia, which had declared sovereignty the previous year.

"This laboratory is the property of the Soviet government," I said. "As its representative, I've made my decision."

We had to find a taxi to get us to the airport for our return flight to Moscow, but I didn't mind.

My trip to Tbilisi exposed me to a more complex problem than graft or medical incompetence. Nationalism in the different Soviet republics was beginning to tear the country apart, and it was tugging at me.

We never studied Kazakh history in school, where even our language was mocked, and over the years I had learned how to integrate myself in the system, how to become a Soviet man. I was now one of the highest-ranking Kazakhs in the Russian army, if not the Soviet government. I knew of only one other senior Kazakh in Moscow, a well-respected general. Kalinin sometimes appeared to be blind to my ethnic features. He would make disparaging comments about Central Asians or people from the Caucasus in my presence as if I were as Russian as he. But when I stepped out of my official car in Moscow, I was often taunted with racial jibes. Nationalist sentiments were rising in Kazakhstan as well as the other Central Asian republics. As more republics declared sovereignty or independence, I began to wonder where my allegiances should lie.

When Lithuania declared independence on March 11, 1990, General Volkov of Gosplan called senior representatives from the Ministry of Health and other organizations linked with our program to his office for an urgent meeting.

"We need to know what projects your agencies are supporting in Lithuania, Estonia, and Latvia," he said.

We would be required to suspend them as part of the economic pressure the Kremlin was applying on the Baltic countries.

Biopreparat had several civilian-run facilities in Lithuania. One was the most modern laboratory in the Soviet Union, thanks to my predecessor, General Anatoly Vorobyov, who so much enjoyed traveling to the Baltics that he funneled over $10 million in hard currency to purchase sophisticated Western equipment.

The lab in Vilnius was the only facility in the country that used genetic engineering techniques to produce interferon, part of the body's natural immune system and used for treatment of hepatitis B and several types of cancer. If it closed down, our top Party officials would lose some of the high-quality medical care they expected.

The order to cut off funding was given and then reversed. Even our political leaders seemed unable to stick to their decisions.

Doubt and uncertainty were creeping into every level of national life. New publications, new revelations, new movies, new books challenged our assumptions every month.

One book in particular created a powerful sensation at Samokatnaya Street—a fictionalized account of the Lysenko genetics controversy that had sent so many scientists to prison in the 1940s and 1950s. It was called *Belye Odezhdy* (*White Robes*). No one had dared to discuss the subject in print before. The book, written by popular historical novelist Vladimir Dudintsev, appeared in 1988, but copies were hard to get. When I finally obtained one from a friend at the office, I stayed up all night to read it and then read it seven or eight times more. Soon all of us were discussing its provocative theme: the role played by the Soviet state and the Communist Party in stifling science.

In April 1990, the government announced that it was planning another reorganization. This time, the Ministry of Medical Industry would be broken up into separate state enterprises. Soon after this announcement, a friend came to me with a job offer from General Yevstigneyev, who had replaced the ailing Lebedinsky as head of the Fifteenth Directorate earlier that year. Yevstigneyev invited me to become his deputy, a position that would mean automatic promotion to the rank of major general.

"Nobody thinks Biopreparat is going to last," my friend warned. "This is probably a safe bet."

I spent the weekend thinking about it. Finally, I turned it down. I had decided to move in a different direction.

———

Kalinin called me into his office one afternoon to discuss the proposed reorganization.

"This could be a way to preserve Biopreparat," he mused. "If we could convince Gorbachev's people to turn us into a separate program again, we could protect ourselves."

"I can't believe Gorbachev will have much time to think about the structure of Biopreparat," I said. "He's got a lot of other things on his mind."

Kalinin looked at me curiously. By now, he knew me well.

"Are you trying to say you have another idea?"

"I do," I said.

"Well, don't sit there dreaming about it."

I drew a deep breath. The announcement of the restructuring plan had given me an opportunity to revive an idea I'd been thinking about ever since Kryuchkov's memo.

"Pasechnik's defection has weakened us and left us vulnerable to American pressure," I said. "We need to find a way to redefine ourselves."

"What are you talking about?"

"If we ask Gorbachev to stop all offensive biological research and production, we'll be in a stronger position to do the pharmaceutical and biodefense work that may become available. Gorbachev won't read a memo about taking Biopreparat out of the ministry, but he might read one that suggests we stop what we're doing. That would be a matter of state policy."

I could see anger rising in Kalinin's face.

"Kryuchkov's memo," he snapped. "I know all about that, and about your little dance with Bykov."

I refused to be cowed.

"If we don't do this," I said, "we won't survive as an agency."

Kalinin said nothing and looked out the window. When he finally spoke, his answer surprised me.

"Go ahead and prepare the kind of memo you're talking about," he said. "If I like what I see, we'll send it up to them."

I went back to my office with a sense of elation. I called in Colonel Pryadkin, who was in charge of long-range planning for

Biopreparat, and dialed General Yevstigneyev at the Fifteenth Directorate.

"I can't believe you, or Kalinin for that matter, would do such a stupid thing," he said. "But if you're going to do it, leave me out of it."

From that point on, Yevstigneyev turned against me. When I met him a few days later at a meeting, he refused to shake my hand.

"Here comes our peacemaker," he said to the officer he was talking to, and turned away.

Only in the Soviet Union could that be an insult.

Pryadkin and I finally managed to draft a decree for Gorbachev to sign. There were just four paragraphs. The first announced that Biopreparat would cease to function as an offensive warfare agency. The final paragraph declared that the agency would be separated from the Ministry of Medical Industry.

Kalinin questioned every word as if in a courtroom.

"All right," he finally said. "Leave it with me. I'll get this to the Kremlin."

For the next several weeks we waited in suspense. Kalinin called Gorbachev's office every day and spoke to one of his assistants whom he knew well, a man named Galkin.

"I don't know why they're taking so long," Kalinin complained. "Galkin keeps telling me there are dozens of papers arriving every day now, and he doesn't know how he can get Gorbachev to see it."

On May 5, 1990, I was called into Kalinin's office. He was smiling and holding up a sheet of paper. Davydov was with him, smiling too.

"We've got it," Kalinin said.

I went over to his desk to read the decree.

Then I went numb. Every paragraph I had drafted was there, but an additional one had been tacked on at the end. It instructed Biopreparat "to organize the necessary work to keep all of its facilities prepared for further manufacture and development."

The first part of the document had ended Biopreparat's function as a biological warfare organization. The last part resurrected it.

I turned furiously to Davydov.

"Volodya, did you do this?" I said.

He didn't answer.

"How can we stop offensive biological research if we have to keep our facilities ready for production?" I demanded.

Kalinin made a dismissive flutter with his hand.

"Look, Kanatjan, you're taking this much too seriously," he said. "With this paper, everyone gets to do what he wants to do."

I wasn't sure whether Kalinin was serious, but I decided to take him at his word. Using the first part of the decree as my authority, I sent a cryptogram to Stepnogorsk and ordered the destruction of the explosive chamber I had devoted so much time and energy to erecting.

Gennady Lepyoshkin, the director of Stepnogorsk, called me as soon as he received the message.

"Have you been drinking, Kanatjan?" he said. "What's gone wrong with your mind?"

"Just do what you are asked to do," I said.

I waited several days, but there was still no word that efforts to dismantle the chamber were under way. I fired off another cryptogram.

"If you don't follow the order," the message said, "you will be fired."

Work started the following week.

Sandakchiev, at Vector, took the news much better. We discussed ways of converting some of the largest buildings to civilian facilities. Thinking of Lithuania, I said I would try to get money so that they could produce interferon.

I went to Siberia several times to oversee the conversion, which was completed by the end of 1990.

Sandakchiev knew how to play politics as well as anyone in our organization. While he was willing to divest some of his biological warfare work, he knew that if he ended it completely he would lose funding from the army. He had also learned—from Kalinin or

Davydov, I imagined—of the extra paragraph in the decree admonishing us to maintain a state of readiness. Sandakchiev owed his loyalties, and his job, to Kalinin. I found out later that the construction of a new building for the cultivation of pathogenic viruses went ahead as planned.

Similar double games were being played around The System. While I closed production lines down, Davydov was authorizing new railcars for the mobile deployment of biological production plants. He could only have done this with Kalinin's encouragement.

The memo was never sent to institute directors. They knew of its existence but were not in a position to act on it without receiving an order from headquarters.

In July 1990, Communist Party organizations in all government agencies were ordered to hold elections for senior management personnel. The election of cadres policy was part of a new campaign to democratize The System. A year earlier, the first elected political assembly in Russia since 1918 gathered to form the Congress of People's Deputies. In February, the Party gave up its seventy-year monopoly on political power.

Mikhail Ladygin, a loyal Party worker in charge of the Communist organization at Samokatnaya Street, asked for my help in setting up the elections.

"You should go to Kalinin," I said.

"I already have. He wants nothing to do with it."

Kalinin believed there was no place for democracy in Biopreparat. It was a military unit, and military discipline had to prevail. Nevertheless, he was too good a politician to stay for long on the wrong side of a Party decision.

We came up with a compromise. Instead of an election, we would hold a "poll" rating candidates on the basis of scientific knowledge and leadership abilities. The poll wouldn't be binding, but Kalinin fully expected to win. To protect himself against surprises, he ordered the list of candidates to be limited to three people: himself, Colonel Davydov, and me.

Ladygin wasn't happy, but he went along. On the appointed

day every employee at headquarters dutifully filled out a question-
naire grading each of us on a scale of one hundred points. The
highest score would win.

When the results came in, Kalinin was not pleased.

I won with an average score of eighty-five points. Kalinin was
narrowly behind with eighty-three points, and Davydov got thirty.

The results didn't thrill me. Even if I had wanted to replace
Kalinin, an "election" was not going to get me the job. It would
take more than superficial reforms to change the way the network
of military and Party insiders anointed and destroyed its leaders.

When Ladygin presented the poll results to a small group in
Kalinin's office and asked when he could make them public, the
general scowled. The blow couldn't have come at a worse time: he
had not yet told Bykov about the May 5 decree freeing Biopreparat
from the medical ministry. Bykov might well use the poll results to
relieve him of his post.

"We don't need to publish the results," I said. "Why not let
people know informally? We can address this again after vacation."

Kalinin eagerly assented. I thought I had removed what might
have been a new source of tension between us, but I was wrong.

Several days later we met with senior management staff for our
last major project review before Kalinin's departure for the summer holidays. Kalinin and I were sitting next to each other in our
usual places at the head of the conference table. When someone
suggested that we review an issue discussed the previous week, I
said, "We don't think that's really necessary."

Kalinin glared at me.

"Are you thinking of yourself in the plural now?" he snapped.

Everyone in the room watched us carefully.

"Of course not," I said. "This was our common decision."

We had reached a breaking point. It was time for me to leave
Samokatnaya Street.

15

VISITORS

Moscow, 1991

In the fall of 1990, I told Kalinin I wanted a new job. He was less upset than he might have been.

One of the largest Biopreparat facilities in Moscow was a scientific conglomerate known as Biomash. Based at the Institute of Applied Biochemistry, with branches in four other cities, it designed and produced most of the basic technical equipment for our weapons plants, ranging from fermenters to concentration and testing equipment.

For several months Kalinin had been trying to get rid of the Biomash director, but he was unable to find a satisfactory replacement.

"Give the post to me," I said in a meeting with him after we had all returned from summer holidays.

"It's a boring job."

"Not to me."

Kalinin had made his career by eliminating threatening adversaries, and now his newest rival was giving up without a fight. Still, the timing of my departure was inconvenient. He had had no time to think of anyone to take my place.

"I need you here," he said.

We worked out a compromise. I agreed to retain my position as first deputy director of Biopreparat and to spend every morning at Samokatnaya Street on administrative and supervisory duties in return for a concurrent appointment to Biomash. This had never been done before, but times were changing. It was a perfect solution for both of us. Kalinin could keep a potential rival safely out of his orbit while continuing to tap the scientific expertise he needed to keep his research program alive. And I could begin to separate myself from a program I no longer considered viable.

Biomash was only fifteen minutes' drive from my apartment in northern Moscow. I would get home early enough to spend evenings with my family for the first time in years. Some of the senior managers were military officers, but the department heads were civilian scientists who took a refreshingly relaxed view of their duties. What made Biomash attractive to me was the fact that 40 percent of its output went to hospitals and civilian medical labs. I intended to increase that percentage. Kalinin and I agreed that my new job would start on December 30, 1990.

It wasn't going to be as clean a separation as we had hoped.

A month after this conversation, in October 1990, we were informed that a "trilateral agreement" had been reached between the Soviet Union, the United States, and Great Britain to organize a series of reciprocal visits to suspected biological warfare facilities. The 1972 Biological Weapons Convention had no provisions for inspections, a limitation that continues to frustrate the international arms control community to this day. Visits were to depend on the trust and goodwill of all sides.

That trust, to the extent it had ever existed, was now considerably frayed. We were told that the diplomatic negotiations had been tense and protracted. The mood filtered down through The System. Anxious debates raged inside the Military-Industrial Commission and the Ministry of Defense.

The memo from the Ministry of Foreign Affairs notifying us of the agreement had left it to us to decide which facilities to open. The defense ministry quickly signaled it wanted no part in this charade. "You can show whatever you want in Biopreparat," General Yevstigneyev snapped when I asked for his suggestions. "But no foreigner is getting into our military sites."

Kalinin, sensing a political opportunity, volunteered that Biopreparat act as the host. He ordered me to decide which of our installations could be "sacrificed" in the interests of East-West relations.

The choices were grim. Though some of our explosive test chambers had been dismantled since Gorbachev's decree, large industrial fermenters at Stepnogorsk and Omutninsk furnished unmistakable evidence of our activities.

At the time, Biopreparat controlled about forty facilities in fifteen cities across the Soviet Union. About a dozen were used exclusively for offensive work, but many others mixed civilian and military functions. Judicious exposure of the Western delegation to these dual-function facilities would protect us while demonstrating our good faith.

We decided to open Obolensk, Vector, Lyubychany (a tiny research institute close to Moscow), and Pasechnik's Leningrad institute. The last was the easiest choice: it was the place Western inspectors were most likely to demand to see, and we had already eliminated all evidence of biowarfare research there.

I never believed that we would be able to pull it off. Anyone with basic knowledge of biological weapons would recognize the signs of our activity. After notifying the labs and preparing instructions for the staff at each institute, I headed over to Biomash with relief. Now it was their problem.

On Friday, January 11, 1991, Kalinin called me into his office for a special meeting. Colonel Vladimir Davydov was there when I arrived. He'd taken the most comfortable chair opposite Kalinin's desk, which I took as a sign of favor. Davydov evidently assumed he would soon step into my shoes as first deputy director. I sat on a sofa facing them both.

The general was in his most agreeable mood.

"Kanatjan, I want to ask you a favor," he began. "The American and British delegation is arriving on Monday. I'm too busy to host them and unfortunately so is Vladimir."

He glanced at Davydov, who refused to meet my eyes.

"I know you don't really want to be involved," Kalinin continued, "but you're the only other senior manager who can do the job. Would you mind acting as host?"

"Yes, I would mind," I said stiffly. "I've only just begun my duties at Biomash, and I don't want to be involved in any of this. I don't see how we can prevent them from learning what we've got."

Kalinin was implacable.

"I thought you might say that. Why don't we come up with a compromise? If you can escort our visitors to the first two facilities, Lyubychany and Obolensk, Vladimir should be free by then and he can take the other two. Kanatjan, we need your help."

As my commanding officer, Kalinin could have ordered me to do as he pleased. But he knew that by appealing to my sense of loyalty, he would lower my resistance. Grudgingly, I accepted.

"Good. There will be a bus outside Smolenskaya at seven o'clock on Monday morning. The delegation will be waiting for you there."

The curious thing is that I began to enjoy the prospect of escorting this delegation. I had never met an American or British scientist before, and this would be the first opportunity to meet people in the enemy camp who knew something of our trade. Not that I intended to exchange notes. As a staunch patriot, I would do everything in my power to prevent the foreigners from drawing the logical conclusions.

The fifteen people waiting outside the Soviet Foreign Ministry on Monday, January 15, looked sleepy and cold. It was still dark, and they were shivering despite their comfortable down parkas and fur-lined boots. The fact that I didn't know a single word of English made that first encounter awkward for me, but it didn't seem to bother Savva Yermoshin. He led the KGB contingent that had been surreptitiously added to our welcoming party.

Yermoshin introduced himself eagerly in broken English to each member of the delegation. He whispered to me later that he'd figured out which ones were spies.

Disconcertingly, they knew a lot about us. One asked through a translator why "Biopreparat chief Kalinin" hadn't come to greet them.

"Unfortunately, Mister Kalinin is extremely busy at the moment," I said. "He wishes he could be here with you and asked me to give you his regards."

That was our first lie, but I enjoyed telling it. Kalinin had made a special point of instructing me never to mention his name.

Everyone on the Soviet side wore suits and ties except for me. I wore an old brown sweater, which had served for the past several months as my silent protest against official protocol. Apparently they found this disconcerting.

"That sweater worried us," one of the visitors laughingly admitted to me during a "reunion" in much happier circumstances at my house in Virginia in 1998. "We thought you wore it to conceal some sort of secret equipment."

We piled into the large bus waiting at the curb. Our first stop was the Institute of Immunology at Lyubychany, about ninety minutes' drive south of Moscow. Yermoshin and his security group crowded into a black minivan and followed closely behind as we set off through Moscow's morning rush-hour traffic.

The bus driver had been advised not to hurry. The strategy, carefully worked out over the previous weeks, was to waste as much time as possible during the nonessential parts of the tour, to reduce the "official" segments in which our visitors could exercise their curiosity. We had also notified our institute directors in advance to stock up on supplies of vodka and cognac, in the hope of befuddling them with Russian hospitality.

The driver obeyed his instructions so punctiliously I was convinced we were lost.

Lyubychany was an easy stop. Since its activities were largely confined to theoretical analysis and defense work, there were no suspicious pathogens on hand.

Nevertheless, we refused to give our visitors the benefit of the doubt. The institute director, a scientist named Zavyalov, spent most of the morning talking about his research projects. Then he provided a sumptuous lunch. By the time he was finished, only a few hours remained for the inspection.

We had reserved two days for each visit. On the second day at Lyubychany, I began to challenge behavior that could be construed as threatening. I stopped Chris Davis, the leader of the British group, when he pulled out a tiny tape recorder.

"Not permitted," I said.

He looked puzzled.

"But we were only told we couldn't videotape," he protested.

After wasting some time in negotiations, I magnanimously conceded the point.

When I returned to Samokatnaya Street later that day I described our feints and evasions to a delighted Kalinin. I decided not to bring up the fact that they had asked after him.

"Wonderful," he said. "Keep it up."

Obolensk, our next destination, was more difficult. It would require finesse to explain the heavily insulated buildings and labs, the animal cages in our bacteriological genetic engineering program, and other elements of a biowarfare infrastructure. It would be hard, I thought, to conceal our projects to develop antibiotic-resistant strains of tularemia, plague, brucellosis, glanders, melioidosis, and anthrax and to pass our work off as biodefense.

General Urakov surpassed my expectations. The crusty director of Obolensk became the soul of conviviality and tact. Attendants paraded in and out of the conference room with platters of sandwiches and drinks.

By now, the delegation had figured out our tricks. They refused to touch anything.

"Could we *finally* get to work?" Davis said as Urakov paused for breath during an extended welcoming speech.

The general was undaunted. He told the foreigners that they were free to see whatever they wanted in his compound.

"However, I must warn you that biodefense work involves very hazardous organisms as I'm sure you know," he said. "If you elect to go into some of our labs, we may have to ask you to stay in quarantine for a couple of weeks. Those are our regulations."

This was partly true. We did have regulations requiring visitors to stay under observation, but they were totally unnecessary in this case: I had given orders that weekend for Obolensk and Vector to be totally disinfected. The delegation could have walked through any one of the labs in shirtsleeves.

Our visitors, naturally, were unaware of this. They hesitated.

"So," said Urakov brightly. "What would you like to see?"

They asked for a complete tour of the complex. We got our first rude surprise when Davis pulled out a map and pointed to a large building.

"Take us there," he said, pointing to the structure that contained Obolensk's newest, and biggest, explosive chamber.

Ah, I said to myself, here it comes. I assumed the map was based on satellite reconnaissance photos. Pasechnik couldn't have known about this.

I went back to the main building on the compound while the delegation divided into groups and set off on their tour. It didn't take long for feathers to fly.

The group led by Davis was taken into the first building he had asked to see. His guide was a senior scientist named Petukhov, who told me the story later.

The visitors were allowed to wander in the corridors until they came to a closed door.

"What's in here?" Davis asked.

No one answered.

"Could you please open it?"

"We lost the key," Petukhov mumbled. "I'll see if I can find a copy."

While the visitors waited impatiently, Petukhov took his time finding a "new" key. He eventually opened the door, but the room was dark.

"Can you turn the light on?" Davis asked in irritation.

"Not possible," Petukhov said. "The bulb is out."

Undeterred, Davis walked right past him and pulled out a flashlight. At that moment, the façade of international cooperation ended.

Petukhov lunged for the flashlight. Davis shouted. The two men tussled back and forth until someone suggested that they take the dispute back to the conference room, where I was awaiting their return.

"What's wrong?" I asked through my interpreter as they stormed in.

"Nothing," raged Davis, "except that when I tried to use a flashlight to look at a room, this KGB guy tried to grab it from me."

"Really!" I said, quite offended. "That man is a very respected scientist. He is not a member of the KGB."

But I conceded that flashlights were not forbidden.

When Davis finally went back and turned the flashlight on, he saw that the walls near the door were dented and pitted—the telltale marks left by fragments of explosives.

"You have been using explosives here," he declared.

"No, no," Petukhov insisted. "Those marks come from the hammers we had to use to make the door fit when we were constructing the building. It was poorly made by our factory, you see."

It was quick thinking, but the wrong answer. We had prepared everyone with a better explanation. They were to say that it was true, explosives were used, but only to test aerosols for defense work. How could we protect our soldiers unless we knew how biological aerosols behaved?

It probably didn't matter that Petukhov forgot his script. They would have drawn the right conclusion regardless, but it was a matter of pride for us to find answers that sounded at least reasonably intelligent.

That evening, we dined at the Obolensk executive dining room. Bottles of cognac were placed at each table, but the plan to get our visitors drunk fell flat.

"I have to tell you we don't believe a word any of you are saying," Davis remarked candidly to me over dinner, speaking through an interpreter. "We know everything."

"I don't know what you mean," I said with as much surprise as I could manufacture. "But you are free to believe what you want."

After the two days at Obolensk, Urakov telephoned Kalinin.

"Kanatjan and I were a great team," he bragged. "We didn't give anything away."

The visit to Obolensk was supposed to conclude my duties as host. But I wasn't surprised when Kalinin told me that Davydov still had "important work" to do.

"I know you don't believe me," Davydov insisted when we met in Kalinin's office. "But I have to go to Irkutsk to check on the new facility for single-cell protein."

"I know that project," I said. "And I know that there's nothing urgent about it."

Kalinin intervened.

"There's no point in arguing," he said evenly. "You've done such a good job, our guests will wonder why you're not staying with them."

I could read what was on Kalinin's mind as clearly as if he'd said it aloud. If the visit continued with no real problems, Biopreparat would look good. But if something went wrong, they could always blame me. Kalinin and Davydov knew how much I now opposed continuing the program. I was clearly dispensable.

I shrugged and went home to pack for Siberia.

We asked Gosplan for special funds to charter a plane to Koltsovo, where Sandakchiev was preparing to welcome us to the Vector base. The charter was worth the extra expense. We wanted to keep our tiny group under scrutiny as we headed into the secret military regions of Siberia.

We left Moscow in the late afternoon and encountered rough weather just as we crossed the Urals, forcing us to make an emergency stop in Sverdlovsk.

Such delays were normal during the Siberian winter, when passengers on scheduled flights sometimes found themselves stranded for days at a time. Having our own plane gave us an edge, but our guests didn't see it that way. They decided this was another example of Soviet duplicity. I would have liked to explain to them that Russian weather respected neither Communists nor capitalists, but I held my tongue.

As soon as they were told where we'd landed, they grew nervous about being trapped in the place where the 1979 anthrax accident had first aroused Western suspicions.

"We won't be here long," I insisted. "We've got clearance to take off first as soon as Novosibirsk opens up."

A few of them ventured out of the VIP lounge, where we'd

parked for the night, into the main terminal, only to withdraw in horror when they saw the hordes of stranded travelers slumped over luggage and sleeping on the floor. This helped convince them that our delay was not staged. When the pilot came in at 4:00 A.M. to announce that we were ready to leave, they followed him outside with relief.

We crossed the icy tarmac and waited for the pilot to kick open the door of the aircraft (it had frozen shut during our layover) before following him inside. Our security forces had not been idle during the stopover. As we left the VIP lounge, one woman in the group complained that her purse was missing.

"It must have been an airport employee," I said, although I doubted any airport worker would have dared to trifle with a group so clearly under KGB protection.

At Koltsovo, a member of the delegation discovered his luggage had been tampered with. When I confronted Yermoshin, he drew back in mock affront, full of protests.

Sandakchiev was waiting for us in the predawn darkness at Novosibirsk with a fleet of Volgas and minivans. He proved an even better host than Urakov. He was genuinely excited by the opportunity to share scientific research with his Western peers.

They didn't share his enthusiasm. As soon as the effusive Armenian launched into a lecture about the latest advances in Soviet immunology, he was cut off.

"Please," one of the visitors said edgily, "we would really like to see your labs."

Sandakchiev looked disappointed. Our guests were treated to another warning about quarantine and then escorted into the facility.

I could see their eyes widen with astonishment as we took them past enormous steel fermenters, larger than what any Western pharmaceutical firm would ever use for the mass-production of vaccines. We had taken them to one of Vector's principal research labs—but we made sure to keep them on the ground floor.

In the upper levels were floors dedicated to work on smallpox, Ebola, Machupo, Marburg, Junin, and other hemorrhagic fevers, as well as VEE, Russian spring and summer encephalitis, and a number of other deadly viruses. Fortunately, the fuzzy protocol of the visit had left them uncertain as to how hard to press us.

They asked whether they could take air samples and smears from some of the lab areas.

"We haven't tried to hide the fact that we work with dangerous strains—for defensive purposes," I replied. "But I don't have any instructions about allowing you to take samples outside the country. We don't want to be responsible for a terrible accident."

Neither did they.

"I can always ask for permission," I added helpfully. "It might take time and you'd have to wait here for an answer, but I'm sure you'll be well treated."

They didn't press further.

Sandakchiev and I noted with relief that they had not brought special equipment. We had feared for months that the visitors would be carrying advanced monitors capable of detecting viral DNA. Such monitors would have picked up irrefutable evidence of smallpox, and we would have a lot of explaining to do.

The only member of our team who truly seemed to be enjoying himself was Yermoshin. He was convinced he'd identified the lead American intelligence officer and spent the rest of our excursion cordially attempting to trip him up.

"He speaks Russian, and he clearly knows as little about biology as I do," Savva whispered to me with delight. "All he cares about is asking everyone political questions."

By the time we got to Leningrad, I had begun to relax. The worst was behind us. Nothing at Pasechnik's old institute would pose a threat. Or so I thought.

During the tour, one of the visitors stopped before an imposing machine.

"What's this?"

I groaned inwardly. I had forgotten about Pasechnik's jet-stream milling equipment. It had been too heavy to move. No one had informed me that the machine was still active, and I silently raged against Pasechnik.

The institute's deputy director, a man named Vinogradov, came up with a quick-witted response.

"For salt," he said. "That's where we mill salt."

I think by then our guests had had enough. They didn't even bother to smile.

On the last night of their stay we gave them a banquet at their hotel. I stood up to offer a toast.

"I know you think we weren't very open," I said. "But please remember this was a first for all of us, after so many years of mistrust between our countries."

I went on, pausing meaningfully, to say, "We all have our secrets . . . in biodefense, but after all, this won't be your last visit, and we look forward to being your guests soon. Once relations improve, things can only get better."

I was rather proud of the speech. I thought I had struck just the right note of candor and diplomatic evasiveness. The day before, coalition forces led by the Americans had launched Operation Desert Storm in Kuwait. I decided to sweeten the moment with an expression of solidarity.

"I would like you to know that a lot of Soviet people support your actions in Iraq," I said. "I truly hope you win."

Strangely, no one reacted. I wondered if our translator had been doing his job.

"Kanatjan," Yermoshin advised me quietly after dinner, "I think you should stay away from politics."

Two weeks after the delegation left, Biopreparat prepared a report for the Military-Industrial Commission. We claimed a kind of victory. Although the delegation had seen enough to make them suspicious, they could prove nothing, and we had given nothing away.

Kalinin was almost as happy with me as he'd been when I developed the tularemia weapon at Omutninsk.

I returned to my office at Biomash, thinking I had earned the right to proceed with my plan to convert at least one corner of the vast Biopreparat empire into an outpost of useful activity.

Throughout that spring and summer the Soviet Union sank into further political disarray. The scientists under my command seemed delighted to be doing peaceful work. One team took the job of converting the mobile production assembly lines used to fill bomblets with biological agents into automated lines for vaccines.

I spent less and less time at Samokatnaya Street. I would occasionally stop in to say hello to old friends, but I kept as far as possible from my second-floor office.

From time to time, I would receive a call from an irritated Kalinin.

"I tried to find you the other day to get you to a meeting at the Central Committee," he said, "but you're always unavailable. Don't imagine you're fooling me. I know what you're up to."

By now, I really didn't care. Eventually he stopped inviting me to the "urgent" meetings that had once been such an important part of my life.

One reason why Kalinin could do little more than complain about the changes I was making at Biomash was that it was now official state policy to convert military plants, whenever possible, to civilian purposes. Often the transformation was ludicrous. Workers at a plant that formerly manufactured jet fighters in the center of Moscow were suddenly producing washing machines and food mixers. Their products were of such poor quality that it was hard to imagine how they would attract even the most hard-pressed Soviet consumer.

All the same, our militarized economy was undeniably changing. A few other civilian managers at Biopreparat headquarters had left the organization completely.

I learned that Kalinin was having trouble keeping some of his large offensive research projects afloat. He reduced my institute's portion of budget allocations, insisting that all managers had to tighten their belts.

The lack of money forced me to look outside for help. Valery Popov, a friend from Biopreparat who had left to become president of the newly formed Russian Biomedical and Pharmaceutical Association, offered to help arrange the financing of some of my projects.

In the summer, Popov introduced me to an American businessman named Joel Taylor, a retired arms-company executive from Austin, Texas, who ran a company called Cornucopia. Taylor had developed a plan to ship used American hospital equipment to Russia, but he couldn't find anyone to supply transportation.

I called friends at the Ministry of Defense who said they would happily supply a cargo plane if someone paid the $30,000 in estimated fuel costs. Popov and I managed to obtain part of the money from private sources in Moscow. After weeks of lobbying, we got a tentative expression of interest in paying the rest of the bill from the Ministry of Health.

"I've arranged for the minister to meet Joel Taylor," Popov announced with excitement one day. "Can you join us?"

I gladly agreed. The meeting was scheduled for August 19.

16

THREE DAYS

Moscow, August 1991

On the morning of the day the Soviet Union began its final passage into history, I had an appointment with my doctor. I was dressed and preparing to leave home when the telephone rang. It was seven o'clock. Joel Taylor's secretary apologized for calling so early.

"Are you still planning to go to the health ministry for your meeting?" she said.

"Of course," I answered testily. "Why shouldn't I be?"

"Don't you know what's going on?"

"No."

"Turn on your television," she said. "I'll call you back in a few minutes."

I turned on the TV. A ballerina was pirouetting in *Swan Lake*. The same grainy version of Tchaikovsky's ballet was on every channel. I wondered why no one had bothered to obtain a better print of the movie: they had shown the same version when Brezhnev had died, the unmistakable signal that a major state event had taken place.

An announcer came on the air, speaking in an arch Soviet verbiage that had been absent from our newscasts for months. A Committee for the State of Emergency had just been formed, she said. Soviet citizens were asked to remember their duties to the Motherland and stay calm. The ballet resumed.

Taylor's secretary called back.

"What's going on?" I asked.

The bulletins had started at six o'clock that morning. Gorbachev had fallen sick at his state dacha in Crimea, where he was on his annual holiday. He had handed over power "temporarily" to the GKChP, she said, using the Russian initials for the emergency committee.

"I'll still be there," I said, and hung up.

I sat back on my bed, too angry to speak. I'd been following the political events with interest during that surreal summer. In late July, President Bush had come to Moscow for another summit with Gorbachev. On August 2, Gorbachev announced his intention to sign a treaty granting the Soviet republics startling new powers, including the right to collect taxes—the most radical change in the federal structure of the Soviet Union in decades. He left on August 4 with his family for Crimea. The official signing ceremony for the treaty was to be held on August 20, the day he planned to return from holiday. This was the nineteenth, and Gorbachev would not be coming back tomorrow—if ever.

Lena was propped up in bed next to me, watching the TV with a fixed gaze. Another announcer was reading "Order Number One" of the GKChP. We were informed that all government institutions had been placed under the authority of the emergency committee. Political parties, strikes, and demonstrations were banned.

The names of the committee members stunned me at first, and then they seemed infuriatingly logical.

Gennady Yanayev, the pasty-faced bureaucrat whom Gorbachev had appointed vice president earlier that year, was the acting president. He was joined by Defense Minister Marshal Dmitri Yazov, a beefy general whom Gorbachev had plucked from the ranks to shake up the armed forces in 1987, and KGB chairman Vladimir Kryuchkov, who had recommended the closure of our biological warfare program.

The other conspirators were Anatoly Lukyanov, chairman of the Supreme Soviet and one of Gorbachev's oldest friends; Oleg Baklanov, chief of military industries at the Central Committee; Boris Pugo, minister of internal affairs; Valentin Pavlov, the prime minister; Alexander Tizyakov, president of the Association of State Enterprise; and Vasily Starodubtsev, head of the Union of Collective Farm Chairmen.

It was a collection of nonentities from a Soviet Union that had seemed to be on its way out: men of the type I had dealt with throughout my career. Baklanov was closely connected with the inner circle that managed Biopreparat. I had no doubt in my mind that they represented a collective disaster.

"Do you think Gorbachev is really sick?" Lena asked me.

"About as sick as I am," I said.

Slava was waiting outside in the Volga. He took me to the Army General Staff Hospital, but we said little.

Later that morning, at Biomash, hardly anyone spoke. People walked past me with their heads down. Our institute's Communist Party chief was pacing in front of my office. He gave me a meaningful smile.

Only the previous month, I had ordered him to move his papers and staff out of the building, in line with a decree from Russian president Boris Yeltsin banning Party cells from government agencies in Russia.

"What do you want?" I said.

"Well, you've heard the news, haven't you?" he said. "We've won!"

"Who is 'we'?"

"The Party . . . normal government," he said with enthusiasm. "We're ready to move back into the building whenever you are."

"No you won't," I said.

The smile died on his face. "What?"

"You will never come back. This is what Yeltsin ordered, and he is the president of Russia."

"You're going to regret this," he said menacingly.

"Get out," I said. "Go to hell."

As soon as he left, most of my senior staff crowded in. One or two who had heard the argument shook my hand, but our conversations were hesitant and stilted. We discussed the morning's events as if they had happened somewhere else. There was a wariness in the room; people were silently choosing sides.

I met Joel Taylor at 1:00 P.M., but the minister never showed. After half an hour of shared pleasantries, I told our translator to advise the American to go home.

"Tell him nobody knows what's going to happen here now," I said.

As I headed back to the car, someone mentioned that crowds were gathering at the White House, the seat of the Russian government.

Kalinin called later that afternoon and asked me to come to Samokatnaya Street. I asked Slava to take an indirect route along the Krasnopresnenskaya Embankment, leading past the White House.

For the first time that day, Slava smiled.

As we approached the Russian parliament building, the crowds grew so large that our car was forced to stop. I got out and began to walk. A few curses flew toward me from people who had seen me emerge from the official limousine, but I plodded on. At the side streets leading up to the parliament entrances, the crowd had piled bags stuffed with garbage to serve as a makeshift barricade. The towering white building was engulfed by a human sea. Throngs of people filled the space between it and the Moscow River, and even more were pouring into the square behind. Some had brought blankets and bags of food.

I wandered among them for a half hour or so and then threaded my way back. Slava was hovering protectively near the car.

"They say Yeltsin has asked everybody in Moscow to come to the White House," he told me. "Some of these people plan to stay here all night."

Later estimates suggested that there were twenty-five thousand government supporters at the White House that afternoon. The number would soon swell to more than three times that size.

———

Kalinin's office was filled with senior managers, including Yer-moshin, Davydov, and a few chiefs of the directorates. They were huddled in conversation.

Davydov, who was closest to the door, grabbed my arm as I walked in. His face glistened with perspiration.

"Kanatjan!" he said giddily. "Isn't this wonderful!"

"Isn't what wonderful?"

"They've finally arrested that idiot Gorbachev! The man who was destroying our country! He should be hung!"

I was still wrapped in the glow of what I had seen at the White House. As I scanned the room, I was struck by the fact that no one seemed particularly surprised by the coup. Davydov's word—"fi-nally"—had an ominous ring.

Kalinin never hid his sympathies for critics of Gorbachev. In his circle of senior bureaucrats and military chiefs, spiteful comments had been circulating for months. Bykov, our minister, used to say "pluralism"—one of the watchwords of the Gorbachev era—as if he were spitting. Muttered threats about "throwing that bastard out" had become so common that I no longer paid attention. Kalinin would have been warned by his friends of an impending action: we were a military organization, and Biopreparat would have been placed on alert.

I was no longer trusted, but Davydov had always been reliable. His beaming face suddenly seemed abominable to me. For the first time in my career, I lost my temper. The room went still.

"I can't wait to hear what you say when Gorbachev comes back!" I shouted at Davydov, who stepped back as if I'd struck him.

Kalinin stood up.

"Kanatjan, calm yourself," he ordered. "There's no reason to get upset. Gorbachev is very sick. You can believe me."

"Tell that to the people at the White House," I said. "Maybe you can convince them to take down their barricades."

"What barricades?" someone asked.

I described the scene on the embankment. Kalinin shook his head. Everyone else looked fixedly at me.

"If only people would understand," he said. "The country is in very good hands. There is nothing to be alarmed about."

"I think there is a lot to be alarmed about!" I was getting worked up again.

Kalinin sighed and looked at his watch.

"Maybe it's time for all of us to go home," he said. "We'll meet here tomorrow and discuss this once more."

As we walked out I asked Yermoshin what they had been talking about.

"We were trying to decide whether to draft a letter of support to the GKChP," Yermoshin said. "Kalinin said it was our duty." Then he added, in a whisper, "Are there really barricades?"

"Go see them," I snapped. Yermoshin was unfazed.

"I know how angry you are," he said. "But I'll tell you something that will make you angrier. Kalinin called Urakov first thing this morning at Obolensk and told him to organize a proclamation of support for the emergency committee. Urakov whipped his officers into line, and he had the proclamation ready by noon. I'm surprised he didn't ask you to do the same thing."

Yermoshin didn't appear at the office for the next two days. He had decided to call in sick, he told me later, to avoid getting "stupid orders."

I thought I was completely alone.

Back in my office, I quickly wrote out a letter of resignation from the Communist Party. I walked down the corridor to the office of Biopreparat's Communist Party organization. Kalinin had allowed it to remain, despite Yeltsin's decree.

The Party man was delighted to see me.

"Don't worry, Kanatjan," he said. "You've got nothing to worry about. You're all paid up."

I stopped in confusion.

"What?"

"Since early this morning, everybody has been coming in here to pay their back dues to the Party," he said, with a touch of sarcasm. "They shirk their obligations for months, and now they see the error of their ways. I've checked your records, and you're one of the few in good standing."

I handed him my letter. His face fell.

"Resign? Are you out of your mind?" he said.

Later that night, as Lena and I were preparing for bed, we heard a noise in the distance, of metal scrunching over pavement. It was the sound of tanks coming from the army base just north of Moscow.

The next morning, Slava looked grim.

"Did you hear the tanks?" he asked.

"Yes," I said.

"There's a line of them coming up from the south as well," he went on woodenly. "They're heading for the White House."

We didn't know at the time that the first tank battalion to reach the Russian parliament that morning had already swiveled its gun barrels in the direction of the Kremlin. The commanders had decided to defend the White House, not attack it.

When I arrived at Biomash, everyone was talking about the televised press conference at which the coup leaders had made their first appearance. They insisted everything was perfectly normal and promised that troops would maintain order in the capital. Despite the censorship clamped on the media, we watched Russian reporters stand up to accuse the self-appointed saviors of our nation of mounting an illegal coup. The cameras focused on Yanayev, whose trembling hands gave away the fact that he had spent most of the day drinking. Our "acting president" seemed to personify the bumbling of his fellow plotters. Through incompetence or oversight, they had failed to arrest Boris Yeltsin and other leading opposition figures, who had now taken sanctuary at the White House.

Yet the comic-opera character of the affair was not comforting. These men were capable of desperate measures. Rumor had it that an attack would be mounted on the parliament buildings that night.

In my office, I wrote out two more letters of resignation. The first was a note resigning from the army. In the second I announced my intention to quit Biopreparat. I put them in separate envelopes and asked Slava to deliver them to Samokatnaya Street.

I had made my decision soon after I heard the tanks in our

neighborhood. Lena didn't try to change my mind, but when I told her I wanted to go to the White House she lost her temper. She told me to think of my children. Then she began to cry.

I was sipping my morning cup of tea, trying to figure out what to do, when a delegation from various departments walked into my office.

"We want to know what you heard from headquarters," one said. I told them, briefly, about my session with Kalinin. I also told them about the proclamation issued by the Obolensk institute.

"We should issue our own proclamation," said the chief of one of our research labs, a man in his fifties. "We need to support democracy."

I looked around the office and saw heads nodding in agreement.

"If it's to be written in the name of our institute," I said, "we should first discuss it in a meeting of the entire staff. Everyone should have a chance to speak his mind."

At three o'clock that afternoon, more than four hundred people jammed into the room where we held our scientific conferences. There weren't enough seats, so some sat on the floor. Others perched on tabletops, fanning themselves with manila folders in the heat. Scanning their anxious faces, I wondered if similar scenes were taking place at that moment in government offices across Moscow.

I stood up, and the buzz of conversation died away.

"I am not here to influence anyone," I began. "I can't speak to you in my capacity as the director of the institute, only as a citizen of the Soviet Union.

"And as a citizen of the Soviet Union, I am calling what's happened a putsch."

Cheers exploded before I could go any further. Some people stood on tables and raised clenched fists.

I told them that with their approval, I would issue on the institute's behalf a proclamation of support for Gorbachev and Yeltsin and send it to the White House. I then read the letter I'd drafted with the department heads and asked them to register their opinion with a show of hands.

"Who agrees?" I said.

Hands shot up across the room.

"Who does not?"

Two people raised their hands. Their neighbors started to jeer.

"Let them explain why!" I shouted above the din.

One of the dissenters was a scientist for whom I had a great deal of respect. He waited patiently until the booing stopped and then stood up to address his colleagues.

"It's clear to me, at least, that we now have a normal government in this country," he said, his voice breaking with emotion. "If we had let things go any further, we could have broken up into little pieces. My father died in the war to ensure that didn't happen."

When he finished, a few people were nodding their heads in agreement.

"Send the letter!" someone called out.

"Send the letter!" the crowd echoed.

Drivers were sent to deliver copies of our note to the White House, Gorbachev's residence, Biopreparat headquarters, and *Obshchaya Gazeta* (*Common Newspaper*), a new journal edited by reporters who were defying the GKChP censorship.

When I returned to my office, I found a message waiting from Kalinin's secretary. I called her immediately.

"Will you be in your office for the rest of the day?" she asked.

"Yes, why?"

"The director intends to pay you a visit," she said, and hung up.

Kalinin arrived at 5:00 P.M. He was carrying my letters of resignation and a copy of the Biomash proclamation of support. There were dark circles under his eyes and his hair was uncombed. In all the years I had known the general, I had never seen him look so frayed.

"You know"—he smiled weakly, breaking the silence between us—"I really think I would like some tea."

He took the chair across from me and placed my letters and the proclamation down on the desk. When an attendant finally walked in with the tea, he gulped it down greedily.

"Are you feeling all right?" I asked.

"I've had better days," he said in a low voice.

We sat in silence until I grew uncomfortable.

"Why did you come?"

Instead of answering, he put his teacup down and placed both hands on the desk. He seemed to be trying to steady himself.

"Kanatjan," he began at last, "I want to admit something. I have a lot of respect for Gorbachev—you must know that. When all this happened, I didn't know what to do. Last night, I couldn't sleep at all."

He waited for an answer, but when none came he continued.

"The trouble is," he said, "these leaders—Yazov, Pugo, Baklanov—are good people. I know them well. They are decent citizens and they love their country. What is a person supposed to think?"

"I can't tell you what to think," I said. "But who are these leaders? Who elected them?"

"That's not the point!" Kalinin replied with some of his old sharpness, and then he slumped into his chair.

"I'm only trying to tell you that they love their country," he said. "They are patriots like you and me, like all of us."

"General, I've made my decision. You will have to decide too."

Kalinin covered his eyes with his hand. Amazingly, he looked as if he was about to cry.

"You don't understand, Kanatjan, you really don't . . . how difficult . . ." He stopped, unable to continue.

I looked away. I thought I knew Kalinin well enough to know he would never forgive me for witnessing his moment of weakness.

Yet for the next extraordinary hour, we talked as we had never talked before. He told me about the problems he faced, the hurdles he was made to jump by bureaucrats at the Central Committee who wanted his job, by the Military-Industrial Commission, by all of his enemies. This man, whom I had never seen lose his bearings, spoke to me not as a subordinate, but as a confessor.

Then, as abruptly as he had begun, he stopped.

"The thing is, you see . . . these are our people," he said stiffly, trying to regain his composure.

"They are not my people," I answered. "I support our president. Maybe he wasn't elected democratically, but—"

Kalinin cut me short with a wave of his hand.

"I don't want to argue with you, Kanatjan," he said, sighing. "All I want is to suggest a compromise."

He pointed to the two papers left on my desk.

"These . . . are premature," he said. "There's going to be a meeting of the Supreme Soviet on August twenty-sixth; Lukyanov has already announced he will take up the matter with them. Why don't we wait to see what happens before you make any rash moves?"

My sympathy for Kalinin evaporated. The motive of his visit now seemed perfectly clear: he understood the meaning of the barricades at the White House as well as I had. He was shrewd enough to understand that the success of the coup was no longer as certain as it had appeared on the first day. Men like Davydov saw the world in primary colors, but the general was too intelligent to believe in his own rhetoric. Our proclamation of support provided him with a political life raft if Gorbachev returned to power, and he needed me around to make it legitimate.

Still, I couldn't help but feel sorry for the man. He had probably poured out more than he had intended in his confession to me.

"I believe our country needs to be strong," he continued. "I know how you feel about our program, but we can't afford to lose anyone now."

He looked at me and tried to smile.

"I beg you. I beg you. Stay."

I had to think quickly. Whether I resigned now or later would make no difference to anyone but me. If I left now as I had intended, my staff might feel betrayed.

"Okay," I said. "I'll leave both letters of resignation with you. You don't have to act on them if you choose not to. If the Supreme Soviet decides that this so-called emergency committee is the legitimate government of the country, I will expect you to put my resignations into action and I will leave this insane country and go to Kazakhstan. If they agree that this is a putsch and we return to the status quo, I'll stay at Biopreparat."

Kalinin looked relieved. I watched as he pulled himself together and stood up, facing me.

"I would advise you, for your own good, to keep your head down," he said stiffly. "Don't do anything foolish until the Supreme Soviet meeting."

"That's my own business."

Neither of us could have known that, even as we spoke, Lukyanov was reporting to his fellow conspirators his failure to secure a quorum of deputies for the August 26 meeting.

After the coup failed, Kalinin destroyed the letter from Urakov's institute and showed the proclamation from Biomash to everyone he met in the ensuing weeks, boasting that "we" at Biopreparat had always known where our true loyalties lay.

If the emergency committee had somehow managed to preserve power, Kalinin would have been the first to propose that the new government abandon Gorbachev's decree halting our biological weapons production lines. And he probably would have been supported by the new leaders, even Kryuchkov.

Vladimir Lebedinsky died soon after the crisis. The old general who had commanded the Fifteenth Directorate for so many years had been ailing for months. He suffered a stroke during an operation to amputate one of his legs.

I was astonished to see how few people showed up at the funeral. Old army comrades like Kalinin and Urakov stayed away, afraid to be seen at a time when sentiment was turning against hard-liners in the military. Nothing deepened my contempt for Kalinin more than his absence that day.

One officer who did show up was Lebedinsky's successor, General Valentin Yevstigneyev. He stood with his head bowed for a long time over the coffin. However insulting he had been during our quarrel over the program's future, I recognized in him a man who was willing to stand up for what he believed in. There were few such people around in the upper levels of the Soviet bureaucracy during those three days in August.

All other Biopreparat facilities remained silent throughout the coup. My institute and Urakov's were the only ones to take a stand.

Sandakchiev called me from Siberia when he heard about our proclamation.

"Fine work, Kanatjan," he said. "I'm glad you stood up to the bastards."

"Why don't you get your people at Vector to do the same?" I asked him.

"Moscow is so far away!" he laughed. "It's all politics. It's got nothing to do with us, nothing at all."

Early on the third day of the crisis, August 21, I was awakened by a call from a man who introduced himself as the duty officer of the Moscow Military District.

"Are you Colonel Kanatjan Alibekov?" he asked.

"Yes."

He cleared his throat.

"I am calling to advise you that you are subject to arrest," he said.

Lena was still sleeping soundly.

"Why?" I asked.

"The colonel general of the Moscow District has announced that all officers who fail to fulfill their duties are subject to a thirty-day preventative arrest," he said, as if reading from a report.

There had been many military officers in the crowd at the institute the day before. Someone had no doubt felt it his duty to inform the authorities of my speech—undoubtedly the informer had raised his hand with the others in favor of our proclamation.

Forty years ago, or even twenty, there would have been no such phone call—just a sharp knock on my door at 3:00 A.M. Times had changed.

"Thank you for telling me," I said.

"You're welcome," he said cordially.

I didn't believe I was really in danger. The night had passed without the attack on the White House that everyone had feared. Yeltsin's parliament, surrounded by tanks from regiments who had announced their support for the Russian government, had survived.

"Who was that?" Lena said.

I told her, and she sat bolt upright in bed.

"Please be careful," she said. "We have three children."

A gray drizzle was falling as I walked outside. I didn't bother to go to Biomash. Slava, who had insisted on keeping me company and acting as my unofficial bodyguard, drove me to the White House.

The crowd was large and restless. Despite the official news blackout, everyone had bits of information to share. Three young men had been killed the previous night in an encounter with a tank on one of the city's boulevards. It was an accident: the tank crew, who were no older than those they killed, had panicked when they were surrounded by a mob of shouting demonstrators. Radios in the crowd were tuned to the Voice of America, and to Ekho Moskvy, a pirate radio station in Moscow that was broadcasting the endless, defiant speeches of parliamentary deputies and Yeltsin supporters inside the building.

Tanks were parked on a bridge within sight of the White House. Hundreds of people had camped in front of them. Soldiers, mostly young conscripts, had unstrapped their helmets and were flirting with girls in the crowd.

The coup was crumbling before our eyes. In the afternoon, Yeltsin announced that members of the emergency committee were on their way to Vnukovo Airport in southern Moscow.

The crowd roared its delight.

"Let's arrest them now!" someone in the throng shouted.

The plotters were in fact rushing to Crimea, where Gorbachev and his family had been held incommunicado for three days. They were bent on explaining their motives to the man they had betrayed. Another jet containing Yeltsin's designated emissary, Vice President Alexander Rutskoi, took off moments later for the same destination to bring the president home.

The two missions arrived at the same time. Gorbachev refused to meet the men from the Kremlin and returned with Rutskoi. Kryuchkov was a captive at the rear of their plane.

Later that night, the shaken president of the Soviet Union stepped on the tarmac at Moscow's Vnukovo Airport, followed by his wife Raisa. Gorbachev raised his hand in a weak salute and then entered an official car to return to the Kremlin. It was over.

Like thousands of other Muscovites, I went home and slept soundly for the first time in three days.

That night, Boris Pugo blew his brains out with his service pistol just as police stepped up to his door.

I went to see Kalinin the following morning. He stood up and offered me his hand. I took it.

"We can all relax now," he said.

My reply was to ask him curtly what he was going to do about Urakov. Kalinin looked uncomfortable when I told him I knew about the proclamation of support for the emergency committee issued from Obolensk. His eyes grew large when I added that Urakov could only atone for his disgraceful behavior by committing suicide.

Kalinin almost laughed.

"Kanatjan," he said in his most condescending voice. "Don't you think that is a bit drastic?"

"Well," I said, "the least you can do is demand his resignation."

"I'll think about it," Kalinin said, and turned away.

Two days later, Marshal Sergei Akhromeyev, the former chief of staff of the Soviet army and a vocal supporter of the putsch, hanged himself in his office.

Within days of the collapse of the coup, it was clear that Gorbachev was too weak to resume his former authority. Having refused in the first hours of his return from Crimea to denounce the Communist Party, he was forced by Yeltsin into a humiliating public rejection of the ideology that had nurtured his career. On August 25, he resigned his position as general secretary.

Soon afterward, I got an urgent call from Kalinin.

"Kanatjan, you must go to the Central Committee offices immediately," he said. "They want us to help them examine documents."

"Why me?" I asked.

Kalinin was defensive.

"There are things there that could embarrass a lot of people," he said. "You know what they are."

I refused. In the end, Kalinin was forced to go himself.

Over the next week, thousands of Party documents were shredded and burned at Central Committee headquarters. Panicked bureaucrats would have burned them all but for the fact that the smoke might have further incensed the mob of demonstrators surrounding the building. The mob had already pulled down the statue of Felix Dzerzhinsky, founder of the Soviet secret police, from his perch in front of the KGB's Lubyanka headquarters.

I learned later that among the documents destroyed were countless papers linking the Central Committee and the KGB to our most secret biological programs, such as Bonfire and Flute.

This was an inspiration, of sorts.

At Biomash, I asked my department chiefs to open our safes and destroy all the instructions and formulas we kept for the making of biological weapons. They did as they were told. I thought there were at least parts of the program that would be impossible to reconstruct.

I was wrong. Copies of every document we burned were held at Samokatnaya Street, where they remain, so far as I know, to this day.

FORTRESS
AMERICA

17

FORT DETRICK

Frederick, Maryland, December 1991

n early December 1991, Colonel Charles Bailey, deputy com-
mander of the United States Army Medical Research Institute of
Infectious Diseases (USAMRIID) at Fort Detrick in Maryland,
gathered his senior managers together for a secret role-playing ex-
ercise.

He divided the managers into two groups. The first group was
asked to pretend it was a visiting Soviet delegation; the second
would answer their questions. "As Soviets, you will be skeptical of
everything you hear," Bailey told the first group. "You're con-
vinced that we are hiding a biological warfare program."

Turning to the second group, he said, "You've got to come up
with plausible answers."

Within two weeks, a team from the Soviet Union would arrive
at Fort Detrick on the first leg of a tour of biological research fa-
cilities in the United States. Everything connected with the tour
was secret. Only a small group of USAMRIID employees—the se-
nior officers designated as official escorts—were told of the im-
pending arrival of the Soviet mission. There would be no press.

*The Office of the Secretary of Defense, which was organizing the
event, imposed a total news blackout.*

When the names of the members of the Soviet delegation were first
sent to the Foreign Ministry, mine wasn't on the list. Kalinin didn't
want me to go. I knew my actions during the coup had made me
unpopular at Samokatnaya Street, but I was irritated nevertheless.
Few people at Biopreparat were more qualified to detect signs of
an offensive biological weapons program. I knew all the possible
ways such a program could be hidden, having directed our con-
cealment efforts since 1988.

"I thought you weren't interested in these things anymore,"
Kalinin had said archly.

But when I reminded him of my service as unwilling host to the
Western visitors, he reluctantly agreed to put me down as an alternate delegate. When Oleg Ignatiev of the Military-Industrial Commission bowed out because of obligations in Moscow, I found
myself in the delegation as the senior-ranking representative of
Biopreparat.

There were thirteen people in our mission—about the same
number the Americans and British had sent in January. We were an
awkwardly mixed group of scientists, army officers, diplomats,
and spies.

Colonel Nikifor Vasiliev of the Fifteenth Directorate led the
seven-man military contingent, which included an officer from the
Defense Ministry's Arms Control Department and an interpreter.
At least one member of the military group worked, by his own ad-
mission, for the Soviet intelligence agencies. He was a GRU colonel
who warned us to tell anyone who asked that he was a representative of the Ministry of Health.

The Biopreparat group was smaller. Joining me were Grigory
Shcherbakov, chief of our scientific directorate, Lev Sandakchiev
from Vector, and General Nikolai Urakov from Obolensk. Urakov's
participation in the trip was uncomfortable for both of us. He had
studiously ignored me ever since I had suggested that he commit suicide. The Ministry of Foreign Affairs sent just two people.

It was not only curiosity about what our rivals were up to that drew me into the mission. I was no longer as dismissive of America's biological warfare efforts as I had been.

A few weeks before our departure, all members of our delegation were summoned to a special briefing at Soviet army headquarters. Maps and satellite surveillance photos of the United States were spread on a large table in the middle of the room. A tall GRU officer holding a wooden pointer lectured us about the four U.S. sites we were scheduled to visit: USAMRIID headquarters at Fort Detrick; Dugway Proving Ground near Salt Lake City in Utah; Pine Bluff Arsenal in Arkansas; and the Salk Center in Swiftwater, Pennsylvania.

As we stared at the maps, he pointed out suspicious structures. At USAMRIID, there was a large circular building that looked like a test chamber. At Pine Bluff, surveillance photos picked up evidence of the movement of "weapons containers."

I was stunned. Why hadn't I heard about this before? I could only conclude that someone had finally decided to mount an aggressive intelligence operation.

I didn't regret having lobbied so hard to close down our program, but I wondered whether Kalinin's efforts to preserve our capacity for weapons research and production had been justified after all.

It was going to be difficult to interpret what we saw. We had agreed to carry no special monitoring or testing equipment. I remembered with amusement the tussle over Chris Davis's flashlight.

Yet something Kalinin had said on the eve of our departure planted the seeds of doubt in my mind.

"Whatever you see there," he told Shcherbakov, who reported the comment to me, "come back with evidence that the Americans are making weapons."

We landed in Washington on Wednesday evening, December 11, 1991. When we unpacked our bags at the Soviet embassy quarters, we learned that our country had disappeared.

American television reported that an agreement reached several days earlier by the leaders of Russia, Belarus, and Ukraine to form the Commonwealth of Independent States had been ratified by the parliaments of each republic—effectively dismembering the Soviet Union.

"This is awful," said Grigory Berdennikov, our escort from the Foreign Ministry (who would later become deputy foreign minister of the new Russia).

"It is," I agreed. "There's no hope for Gorbachev now."

Berdennikov shook his head.

"You don't understand," he said. "We're carrying passports from an extinct country. The Americans will probably tell us to go home."

Our hosts were either too polite or too guarded to mention the subject the next morning.

We were driven in a large bus through the mist-shrouded farm country of Maryland. All I could see through the windows were indecipherable highway signs and large cars whizzing past at breakneck speeds. When we reached Fort Detrick, I relaxed. The place was reassuringly familiar.

Dozens of brick and concrete buildings were spread across a two-hundred-acre site that had once housed a National Guard airfield and training camp. There were large pipes running beside some of the buildings and a tower that could have been a heating station. The configuration reminded me of a pharmaceutical production plant. We turned off a busy highway lined with gas stations and fast-food establishments to enter the main gate, where a guard waved us through. An animal hospital faced the complex on the other side of the road.

Colonel Ron Williams, commander of Fort Detrick, gave a welcoming speech and then turned the proceedings over to Charles Bailey.

As deputy commander of USAMRIID, Bailey was my American counterpart. He was an easygoing man with sandy hair and a soft Oklahoma drawl. He considered himself a scientist more than a military officer. We had a lot in common, though neither of us knew it when we faced off for the first time across a table at Fort Detrick. Within a few years, we would be colleagues at a biotechnology firm in Virginia.

My first reaction was of acute discomfort: he wouldn't stop smiling at me.

Much later, Bailey told me he had interpreted the scowls I kept darting his way as evidence that I was a spy. I thought he was being disrespectful. The more they smiled, the more we were on our guard.

The Americans gave us a map of the compound and asked us to choose which buildings we wanted to visit. We held a quick caucus. The first building we chose was a large laboratory. Technicians in white coats explained that they were developing antidotes to toxins produced by certain animals and shellfish. They were friendly and open—overly so, for my tastes—answering our questions with such ease that I despaired of ever penetrating beneath the surface. I told our delegation afterward that we would have to use more aggressive tactics.

Back on the bus, Colonel Vasiliev pulled out the map and motioned one of our escorts over to his seat.

"What's this building?" he said, pointing to a circular shape located at one corner of the compound. It was the building identified as a test chamber in our briefing back in Moscow.

The American looked confused. He went over to the other members of his team, holding the map.

"There's nothing there," one said.

I smiled to myself. What fools did these people take us for?

We insisted on being driven to the "nonexistent" building. After twenty minutes, the bus pulled up in front of a tall structure shaped like an upside-down ice cream cone. A pair of bay doors stood open. Through them, we could see a pile of grayish powder.

We told our interpreter to ask Bailey what it was. When he returned, he was smiling.

"He says it's salt."

"Salt?"

"Yes, it's what they use to cover roads in the winter."

Vasiliev was dubious. He went to the pile, stuck his finger in, and then put it to his mouth.

"So?"

He looked embarrassed.

"It's salt," he said.

We visited another lab which, we were told, was dedicated to developing vaccines against biological agents such as anthrax. The small size of the operation made it clear that weapons production was out of the question there. The Americans had just two specialists in anthrax. We had two thousand.

In another building, one member of our military contingent impulsively decided to break protocol. Without warning, he clambered onto a lab bench and, to the horror of our hosts, started to remove tiles from the ceiling. I cringed. We were on the second story of a two-story building. I couldn't imagine what he expected to find. He had caught the Americans off guard. After that incident, Bailey's smile began to fade.

Our suspicions were not entirely unfounded. While we had no clear sense of the present state of the Americans' biological weapons program, we knew what they were capable of. In fact, we knew a lot more than they suspected.

When I collaborated on a secret history of the Soviet and American programs after my defection, I was amazed to discover how closely the research efforts of both countries dovetailed between 1945 and 1969. The same agents, even the same types of aerosols, were used in experiments occurring sometimes less than a year apart.

Bill Patrick, who was in charge of biological weapons development at Fort Detrick until 1969, was my partner in the joint history-writing project. Then in his mid-sixties, Patrick was one of the few Americans I'd met who seemed to understand the technology of bioweaponeering. An accomplished microbiologist with a dry wit, he had been responsible for groundbreaking work on plague and tularemia weapons whose formulations remain classified in government archives. He has since become one of America's foremost experts on biological defense.

Patrick noticed the parallel as well.

"When we worked on something, you seemed to be working on it a short time later," he told me. "It's amazing that two countries so far apart could undertake such similar courses."

The uncanny similarity between our programs may have been more than a coincidence. Pavel Sudoplatov, a former general of the

NKVD (the precursor of the KGB), provided an unwitting clue in his memoirs, published in 1996. Sudoplatov reported almost off-handedly that classified U.S. material on bacteriological weapons research had been transmitted on a regular basis to Moscow during the 1940s and 1950s. He passed these reports to "Laboratory X," which he described as an institute directed by one of the senior members of the Soviet Academy of Sciences.

I immediately recognized Laboratory X as Laboratory 12, the unit operated by the KGB's First Main Directorate, where my friend Valery Butuzov had worked for so many years developing assassination weapons. If such information was being obtained by the KGB, it was more than likely to have been shared with other parts of our program, especially after control of biological weapons research passed from the KGB to the army in the postwar years.

We obtained significant data from material published in American and European scientific journals, but American decisions on which strains of biological agents to research, which nutrient media to use, and which aerosols to develop were highly classified. There had to be at least one informer, if not several, in the American program. Patrick told me that no one he'd worked with had ever suspected the presence of a spy inside the American biological weapons labs. But he agreed the evidence was compelling.

The United States was a relative latecomer to the field of biological weapons. Great Britain and Canada began investigating biological agents and delivery systems in 1940, but President Franklin D. Roosevelt didn't establish a program until fifteen months after the United States entered World War II, in March 1943. Based on what Patrick told me of the early history of their program, Americans knew nothing of the ambitious weapons-making drive we had launched in the 1920s.

Perhaps Washington's lack of knowledge was based on its lack of curiosity. The Americans had been skeptical of the value of biological warfare from the start—a skepticism that continues to influence policymakers today.

In 1941, before Pearl Harbor, Secretary of War Henry Stimson asked Dr. Frank Jewett, then president of the National Academy of

Sciences, to put together a working group to investigate the feasibility of biological warfare. Stimson remained unconvinced after he saw the report. "Biological warfare is dirty business," he wrote to Roosevelt in 1942.

Declaring that the military advantages of germ weapons were "debatable," Stimson conceded that "any method which appears to offer advantages to a nation at war will be vigorously explored by that nation." The Americans were not fully persuaded until their British and Canadian allies noted that the Germans were suspected of having used glanders against Romanian cavalry during World War I and seemed to be amassing a larger bacteriological arsenal. The Canadians had converted an agricultural experiment station in Suffield, Alberta, into a testing area for anthrax. In southern England, an old chemical warfare proving ground at Porton Down was adapted for the same purpose.

In America, a secret biological warfare unit called the War Research Service was created to work with its British and Canadian counterparts. George W. Merck, president of Merck & Co. Inc., a leading U.S. pharmaceutical firm, was the unit's first director. Under his leadership, it soon took the leading role in planning biological weapons research for the Allied war effort.

Merck assembled a "brain trust" of scientists from universities and private industry to identify likely spots for research, production, and testing. They settled on four principal sites: a 2,000-acre tract on Horn Island near Pascagoula, Mississippi; the Dugway chemical warfare testing facility in the Utah desert; a 6,100-acre munitions manufacturing complex at Terre Haute, Indiana; and the old National Guard installation at Frederick, Maryland.

The Maryland site, renamed Camp Detrick, was to be the heart of American biological weapons research. Its purpose was kept as secret as the laboratories in Los Alamos, where Manhattan Project scientists developed the first atomic bomb. More than seventeen hundred people worked at Camp Detrick during the war years, investigating glanders, brucellosis, cholera, dysentery, plague, and typhus.

Anthrax was the largest project. Scientists built a pilot plant capable of producing anthrax in ten-thousand-gallon tanks. They were so successful that Britain placed an order for five hundred thousand anthrax bombs from the facility in September 1944.

None of the weapons developed in America was ever used in World War II. Fears that the Germans would use biological munitions in their unmanned "buzz bomb" raids over English cities, or to repel the D-Day force, never materialized. When hostilities ended in Europe, President Truman briefly flirted with the idea of using anticrop agents and antipersonnel munitions against Japan as an alternative to the new atomic bomb.

Victory in the war left America with an unused arsenal of biological weapons, a vast technological and research base, and a secret network that rivaled its nuclear weapons complex. Some facilities were phased out, but revelations about Japan's Unit 731 forestalled any serious discussion about ending the American program.

Like us, Americans learned about Japan's germ warfare operations from captured documents and prisoners of war. Camp Detrick sent scientists to Japan to interrogate the commanders of Unit 731, who gave details of their program in return for avoiding prosecution for war crimes. Their reports convinced Washington that biological weapons could be developed in greater quantities and with far greater effectiveness than anyone had suspected. The British came to a similar conclusion and decided to upgrade their testing and research unit at Porton Down and at a test site on the Scottish island of Gruinard.

Putting aside their initial skepticism, Americans began a sophisticated program of biological weapons development that would last more than twenty years and intensify an arms race no less threatening than its more well known nuclear counterpart.

Beginning in 1951, agricultural agents were developed at Camp Detrick and other facilities to attack the Soviet wheat crop and the rice paddies of Communist China. The pathogens were stored at Edgewood Arsenal in Maryland, as well as at Rocky Mountain Arsenal near Denver, which also manufactured plutonium for nuclear weapons.

U.S. bioweaponeers went on to explore antipersonnel agents such as tularemia, Venezuelan equine encephalitis, and staphylococcal enterotoxin B. Aerosols were tested on animals at Deseret Island in the Pacific Ocean and at Dugway Proving Ground in Utah. They conducted experiments with simulated weapons, as we did, in urban areas.

Human tests took place in 1955 on a group of young Seventh Day Adventists who volunteered as an alternative to military service. The volunteers were exposed to Q fever, which is not lethal and can be treated with antibiotics, in a program called Project Whitecoat, or Operation CD-22.

By the late 1960s, twenty-two microorganisms were under study, and there were plans to weaponize hemorrhagic fevers such as the Machupo virus and Rift Valley fever. The scientists at Fort Detrick were looking into the possibilities presented by genetic engineering when their program was dealt a fatal blow.

Twenty-five years after a presidential advisory board launched America's experiment with biological warfare, a panel appointed by President Nixon recommended killing it.

American doubts about the military value of biological weapons had never completely disappeared. By the late 1960s, public anger over the development of biological as well as chemical arms had melded into the larger protests against the Vietnam War. Pickets appeared every day at Fort Detrick and other installations around the country. Nixon, convinced by his advisers that biological warfare was impractical, signed an executive order on November 25, 1969, renouncing the use of lethal biological agents and weapons and promising to confine American biological research to "defensive measures" such as immunization and biosafety.

We didn't believe a word of Nixon's announcement. Even though the massive U.S. biological munitions stockpile was ordered to be destroyed, and some twenty-two hundred researchers and technicians lost their jobs, we thought the Americans were only wrapping a thicker cloak around their activities.

Nixon turned most of Camp Detrick's buildings over to the National Cancer Institute and assigned to the complex the task of finding a cure for cancer. This, in Nixon's words, would demonstrate how the United States could "beat swords into plowshares." But we also noted that a small army medical unit had begun work at Fort Detrick. This unit, known as the United States Army Medical Research Institute of Infectious Diseases and ostensibly dedicated to biological defense, seemed to expand in importance and strength each year. Former bioweaponeers like Bill Patrick had gone to work for it. Even if our intelligence activities couldn't come up with concrete evidence of offensive work, there could be no doubt that such work continued.

Press reports and transcripts of congressional hearings indicated that many prominent Americans believed it too. This strengthened our conviction that USAMRIID, like Biopreparat, hid its real purpose from the world. Some American experts charged that the Central Intelligence Agency, which had operated a secret unit inside Camp Detrick since 1952 to investigate "paramilitary" uses of biological weapons, continued to stockpile and develop those agents after 1969. The CIA of course denied this, but we knew the value of intelligence agency denials.

During our first days in America, we felt it would take all our ingenuity to ferret out the truth.

We flew to Salt Lake City, Utah, on a hundred-seat plane provided by Vice President Dan Quayle's office. The good food and seemingly endless supply of liquor made me recall with chagrin our trouble-plagued flight to Siberia the previous year. On our way from the airport, I stared in wonderment at the well-paved highways, the well-stocked stores, and the luxurious homes where ordinary Americans lived.

I didn't share my thoughts with other members of our delegation. Sandakchiev had been to the United States before, and he would have laughed at my naïveté. There was no point in comparing travel impressions with Urakov, and our Defense Ministry comrades were too absorbed in their mission strategy to sightsee.

Colonel Frank Cox, the commander of Dugway, met us on arrival at the proving ground, eighty miles from the Utah capital. With disarming candor, he went over the history of bacteriological and chemical testing at the site, which opened in 1942. Since 1969, he said, no biological weapons had been developed or tested there.

More than six hundred buildings were spread across thousands of acres of desert. Dugway was going to be more challenging than Fort Detrick.

We were taken to a large complex designated as the Life Sciences Lab. It was a compound of ten buildings set against a stark landscape of cactus and tumbleweed, and it immediately set off alarm bells in my mind.

The configuration of the structures matched a part of our Stepnogorsk compound. There were sheds for disinfecting equipment and vehicles for transporting animals, and inside some of the buildings I could see tiny rooms similar to those we used in our sanitary passageways for donning protective suits. The largest building looked like a testing facility and, nearby, were distinctive structures with thick walls and loosely fitted roofs—a telltale sign that they had been used to store explosives. In other buildings at the complex, we saw rooms with equipment similar to that which we used to conduct animal autopsies.

But there were no animals, no cages, not even the footprint of experimental weapons activity. The door fittings on many buildings were rusty and creaked when opened. In some, paint was flaking off the walls. The dozen or so lab technicians who worked at the compound seemed lost inside the vast interiors.

Cox's assistants told us the facility was used to test simulants of biological weapons. The main mission, we were informed, was to explore methods of protecting troops and military equipment from biological and chemical attack. They showed us a lab earmarked for the development of devices to detect the presence of biological agents in the air.

Helicopters ferried us to other sites. Our escorts answered every one of our questions with no apparent hesitation. I was impressed, yet I knew that our own technicians had also been well rehearsed.

"They're doing nothing here," declared Sandakchiev.

Urakov said nothing. The military contingent was annoyed.

As we flew to our next stop in Arkansas, we held anxious, whispered conversations.

"This whole trip is just eyewash," Vasiliev said, who came over to my seat to share a drink. "They're not going to give anything away."

It was true. The Americans were doing a much better job of hiding evidence than I had given them credit for. But my doubts were growing.

The Pine Bluff Arsenal in Arkansas manufactured chemical munitions during World War II. In 1953 the facility expanded to produce biological warfare agents, but the installation was turned over to the U.S. Food and Drug Administration (FDA) in 1969 for civilian research. This, at least, is what we were told by our hosts. To my discomfort, the evidence seemed to support it.

The layout of the Pine Bluff compound was once again similar to our own facilities. One building housed giant grayish blue tanks used to treat contaminated wastes. We had tanks like these in our plants. When our guides unlocked the door and brought us inside, I noticed that the floor was covered with a layer of dust. The tanks were wrapped in insulating material, cracking with age. As I wandered through the building, a black notebook on the floor caught my eye. I bent down to pick it up, blew away the dust, and quickly scanned its pages. I couldn't read the handwriting, but the year in which the entries were made was printed clearly: it was marked 1973.

We went into another facility that had once been used for assembling and filling bomblets with biological agents. It had since been reconfigured and divided into laboratories where American biologists worked alongside cages of mice and other animals.

When we found out what they were doing, the scientists in our group became enthralled. The shortage of space at Pine Bluff had forced the Americans to convert their old weapons plant into a center for medical research into immunosuppressive substances—substances capable of preventing the body from mounting a normal defense against invading bacteria.

This research is of immense importance to organ transplant surgery, as doctors must find ways of preventing the body from rejecting a transplanted heart or kidney. The technicians were busy grafting pieces of bird skin and other organs on the mice.

We spoke with the scientists for several hours, to the evident displeasure of some of the nonscientific military members of our group. Sandakchiev couldn't stop asking questions. By then, I was convinced that the Americans were no longer involved in biological weapons work.

Our military cohorts didn't agree, and the difference in perspective soon became embarrassing. On the second day of our Arkansas visit, I climbed into the bus beside one of the Defense Ministry officers, a colonel named Zukov. As our escorts pointed out various structures passing by our window, I dozed.

Suddenly Zukov began to shout. "Stop the bus! Stop the bus!"

I woke up in alarm.

"What's wrong?"

He pointed to a tall metal structure standing on a rise.

"We have to check that out," he said.

"Don't be ridiculous. It's a water tower."

"I don't think so," he said.

Zukov ran to the water tower. He began to climb it, all the way to the top. Behind me, I could hear our American escorts trying to stifle their laughter. One took a picture.

At that point, the absurdity of our quest was clear to me. We could go on like this for weeks, but it would get us nowhere. Perhaps there were other sites in America where secret biological weapons work was going on, but these were the ones we'd asked to see. I remembered the conviction with which our GRU briefer had spoken of evidence of weapons work and had shown us his surveillance photos.

We were the victims of our own gullibility. I have come to believe that the most senior Soviet officials must have known all along that the Americans had no serious biological warfare program after 1969—after all, our intelligence agencies were among the best at their craft, and they had not come up with any real evidence. But the fiction had been necessary to instill in us a sense of urgency. The Soviet biological warfare program, born initially out of fear and insecurity, had long since become a hostage to Kremlin politics. This would explain why Kryuchkov had been so willing to trade it away in 1990 and why bureaucrats like Kalinin and Bykov refused to give it up.

In the city of Little Rock, thirty-five miles north of Pine Bluff, we got a taste of American politics. Shcherbakov and I were sitting at the Excelsior Hotel bar on the first evening after checking in when we became aware of a crowd streaming past us. Curious, we followed it to a large room adjacent to the lobby. There was a great deal of cheering and waving of signs. A boyish light-haired man stood on a raised platform at the front of the room, raising his hands to acknowledge the applause. Shcherbakov, who knew some English and had a passing acquaintance with U.S. affairs, told me this energetically smiling man had just announced his candidacy for president of the United States.

"He's the governor," confided Shcherbakov, "but he doesn't have a chance. No one ever became president from Arkansas."

Before we departed Pine Bluff, the director handed out diplomas certifying that we were "Arkansas Travelers." They were signed by Governor William Jefferson Clinton.

Our last stop was the Salk Center at Swiftwater in northern Pennsylvania. A research institute for the development of vaccines, it had no military past, present, or future, at least none we could discern. We returned wearily to Washington, D.C., where the approaching Christmas holiday ended all further talk of biological weapons—to the relief both of our hosts and of ourselves.

On our final afternoon in America, we were taken on a tour of the capital. Our guide was Lisa Bronson, an official from the division responsible for disarmament policy in the Department of Defense who had been in Moscow the previous fall to negotiate the terms of our visit. She had accompanied us throughout our journey across America. A sharp, brisk woman in her mid-thirties, she had gotten to know most of us well. At various stops along the way, she had challenged us about the Soviet biological weapons program. Naturally, we denied we had one. But I admired her persistence.

Standing on Pennsylvania Avenue near the White House, we steered the conversation in a different direction.

"What do scientists actually earn here?" someone asked.

There was no interpreter, and Sandakchiev, who spoke English fairly well, translated.

"That depends on your experience," she answered. "A government scientist can make between fifty thousand and seventy thousand dollars, but a scientist in the private sector could earn up to two hundred thousand dollars a year."

We looked at her in astonishment. At the time, a top-level Russian scientist could expect to earn the equivalent of about one hundred dollars a month. I screwed up the courage to ask a question of my own.

"With my experience," I said, "could I find a job here?"

She smiled. "If you know English."

"Okay," I said as Sandakchiev translated. "If I ever come here, I'll ask for your help."

Everyone started to laugh, including me.

18

COMMUNIST PROSPEKT

Almaty, 1992

Gorbachev resigned the day we returned to Moscow, on December 25, 1991. Lena told me the news as I walked into my apartment late that evening, my arms laden with gifts from the United States. On New Year's Eve, the red hammer-and-sickle flag of the Soviet Union came down from the Kremlin. In its place rose the Russian tricolor, the flag that had waved over the Russian White House the previous August.

The new government of Russia seized the imagination of the world. It wasn't, however, my government. I was an officer of a colonial empire that no longer existed, a stranger in a country that was not my own. I was entitled to become a citizen of Russia, but in truth I was now a foreigner.

Tens of thousands of people like me were orphaned by the collapse of the Soviet Union. Whether we were Kazakhs, Ukrainians, Moldovans, or Azerbaijanis, no matter how closely we were linked to Russia by marriage or government position, and however much we welcomed the new climate of freedom, we faced the same difficult choice. Should we go "home" to countries with which we had no real connection, or live as aliens in what would from now on be an adopted homeland?

On January 13, 1992, seventeen years after I had received my junior lieutenant's commission, I left the army. My letter of resignation had been in Kalinin's safe at Samokatnaya Street since the failed coup. He was surprised when I asked him to put the resignation papers through. No sane person, he believed, would voluntarily give up the perks of military rank.

I wasn't ready to break with Russia completely. But I thought severing my military connections would liberate me from a program I had begun to despise. This turned out to be a futile hope.

New leaders assumed control in the army and the KGB, but the power structure in both organizations remained unchanged. The Military-Industrial Commission was attached to the newly formed Russian Ministry of Industry, with its function preserved. One by one, former Soviet institutions merged with the new government, bringing with them the cadres of apparatchiks who had ruled the old empire. The hopes of the new Russian democracy were being compromised behind the scenes even as we were promised a new way of life.

The biological warfare program followed the same path. Biopreparat's production capacity was destroyed under Gorbachev's orders. It should properly have been disbanded, or at least merged with the new state pharmaceutical enterprise, but Kalinin was determined to preserve his autonomy—and he enjoyed quiet support from the military bureaucracy.

Our report on the U.S. visit was crucial to his strategy. If he could demonstrate that America was conducting offensive research, he could convince the Yeltsin government that Biopreparat was essential. Yet massaging the facts couldn't make a biological warfare program out of what we had seen.

I should have realized this wouldn't deter him.

Attached to our ten-page report was a "summary" prepared by Kalinin and Grigory Shcherbakov. It claimed our observations proved the continued existence of an American weapons program. The report was duly sent to the Kremlin along with a further recommendation from the Fifteenth Directorate that Russia's offensive research continue. This was the last straw. I wrote out a second letter of resignation from Biopreparat and took it to headquarters.

Kalinin opened my letter in exaggerated slow motion, fingering it as if it were toxic. When he finished it, he looked up with a puzzled expression.

"What do you think you will do with yourself?" he said.

"I don't know yet. Maybe I'll start a private business. Maybe I'll go to Kazakhstan. It's my motherland, after all."

"Your motherland?" He shook his head. "You swore to serve the Soviet Union, just like me."

"I lived in a country called the Soviet Union," I replied. "I served it loyally. But it doesn't exist anymore. So now I'm free."

Kalinin's face darkened.

"I always suspected you were the type who thought he was too good for Russia," he said.

"You can think what you like," I said, my temper beginning to rise. I had promised myself not to let him provoke me, but I could feel that determination ebbing.

"All right," he said, holding his hand up as a peace signal. "We don't have to fight, but be reasonable. Who around here can take your place?"

"Lots of people want the job. You can make one of them happy."

He stroked his chin and smiled. He had decided to charm me.

"You don't know how valuable you are to me, and to this organization," he said. "Maybe you should give this some more thought."

It was a strange moment. The man with whom I had argued and fought throughout the previous two years, who knew that I hated everything he stood for, now appeared anxious to keep me by his side.

"No," I said. "My decision is final."

"Well, my decision is that I'm not giving you permission to leave."

"You can do what you like, but I'm not under your command anymore," I shot back. "Whether you agree or not, I will be gone by next week."

He stiffened. "Are you giving me an ultimatum? You're the director of an institute. You're not allowed to leave."

"I don't want to work in this program any longer," I said. "Or with you."

He grabbed my letter from his desk and threw it at me.

"You are a traitor!" he shouted. "I always knew you would betray me!"

I threw the letter back.

"I haven't betrayed anyone," I said. "Think about last August before you accuse anyone of betrayal."

I turned and walked out of his office, past a startled Tatyana, who must have heard the entire argument. I went down the corridor to our personnel office and handed in my secret passes and credentials. The building was quiet. Several people poked their heads out of offices as I walked by, but no one said a word.

I walked down the marble staircase and pushed open the door. The KGB guard outside saluted as I walked through the courtyard to get my car. I had driven over in my Zhiguli. I never wanted to see an official Volga for the rest of my life.

The KGB man saluted again as I drove past him out the gate. It started to snow.

I spent the next several days clearing out my office. Kalinin didn't call. I never saw him again.

In that first year after communism it sometimes seemed as if you only had to leave your apartment to make money. Friends had pockets bulging with rubles and dollars. One handed me a sports bag so heavy I could barely lift it. "I've got one hundred thousand dollars inside," he announced proudly. I was unemployed for the first time in twenty years, but poverty was not one of my fears. Government officials everywhere, in and out of office, were regarded as prime catches for the new Russian *biznesmeni*.

Within a few weeks of my departure, I was working as the Moscow representative of a Kazakh bank. My brother had given them my name, and they hired me immediately to develop their overseas interests. I had no aptitude for finance, but I was soon making deals like everyone else.

The idea at the time was to make as many millions as you could, in whatever way you could, before it ended in an inevitable disaster. Corruption and crime were everywhere, and I heard ominous talk of high-flying acquaintances put "on the time-clock" by mobsters who had loaned them money and were now doubling the interest every day that the debt remained unpaid.

My phones soon started to click and crackle every time I made a call. The telephone company insisted that nothing was wrong with the line. The noise vanished when I changed the number, only to reappear after a few days. When I was away on business trips Lena would receive mystifying calls from people who introduced themselves as "general" or "colonel." They would ask when I was expected back. We wouldn't hear from them again.

In the spring of 1992, I placed a call to a business partner during a meeting in my office. Just as I finished dialing his number, I remembered something I'd forgotten to say and put the phone back on the hook. Five minutes later, my partner, Naum, called.

"Kanatjan, something's wrong," he said.

"What's wrong?"

"My phone rang once, and when I picked it up no one was there. But I heard you talking to someone else."

"It's just a bad connection," I said.

"No, there's more. I not only heard what you were saying, I heard everybody else too. It felt like I was in the middle of the room."

Then he repeated word for word everything that had been said in my meeting.

"That's not just a poor connection," he said.

One evening a policeman appeared on the sidewalk outside our building. He was gone the next morning, but a new one showed up when I returned from work that night. From then on police made frequent appearances, jotting down in their notepads my arrivals and departures. They never came when I was out of Moscow.

On April 11, Yeltsin signed a decree banning offensive biological warfare research. I heard about it almost immediately from one of my former colleagues, and I was overjoyed. It meant, or so I thought, that Kalinin had lost his battle. The decree banned all offensive biological work and cut research into defensive programs by 50 percent. The Fifteenth Directorate was dissolved and replaced by a new army department of nuclear, biological, and chemical defense. The decree didn't mention Biopreparat, but I felt as if a huge burden had been lifted from my shoulders. My former life was no longer a military secret. Presumably no one would care what I did with my new one.

A few weeks later, I concluded a sale of oil from Kazakhstan with a business associate whom I'd been dealing with over the previous six months. The client, Mark Severinovsky, was a flamboyant character, a diamond merchant and entrepreneur who enjoyed sprinkling his conversation with the names of cities he'd visited: Tel Aviv, London, Bonn. Our discussion had never gone much beyond business, but after we finished our oil negotiations, we decided to unwind over coffee.

Halfway through our conversation, he leaned back and said, "Kanatjan, I hear that you want to leave the country."

"Who told you that?" I was stunned.

"It doesn't matter."

"And why would you care?"

"You're carrying around a lot of secret information in your head."

I considered what to say next. Finally, I told him that Yeltsin's decree had made the issue academic.

He suggested that "others" would see things differently, that I had no idea how damaging what I knew might be. Damaging to whom? I asked, but he merely smiled and said he was telling me this for my own good and returned to his coffee as if nothing had happened.

The idea of returning to Kazakhstan gradually went from a half-formed notion to a desperate conviction. Kazakhstan had declared independence while I was still on my trip to America, on December 16, 1991, and I considered applying for citizenship.

I was spending at least one week a month in Almaty on business, staying at my parents' apartment. As soon as I walked through the door of the old building on Communist Prospekt where I'd grown up, I escaped from the tensions of my past and present lives. My family still knew nothing about my career. My sister once confided that she thought I was involved in a secret program to clone people.

In Almaty, my mother showed me a newspaper with a decree by President Nursultan Nazarbayev offering citizenship to Kazakhs living outside the country. Nazarbayev especially welcomed scientists, doctors, and engineers, challenging them to participate in the country's transformation. In 1990, when I was still at Biopreparat, I had received a vaguely worded invitation to serve as Kazakh health minister. I had given it little thought, convinced that the burgeoning Soviet democracy would achieve more than the corrupt authoritarian clans of Central Asia. But things had changed.

In June of 1992, I received a phone call at my Moscow office from a man who introduced himself as Mikhail Safrygin, first deputy minister of defense of Kazakhstan.

"Are you by any chance planning a visit to Almaty any time soon?" he asked politely.

"Yes," I said. "I'll be there next week."

"Would you mind stopping by our ministry? We have a job in which you might be interested."

This was the opportunity I'd been waiting for. I didn't expect the job as health minister to come my way again, but the leaders of the new government obviously knew my background in military medicine. I imagined that they needed someone who could organize a health service for the new Kazakh army.

I set out for the interview in a new and expensive suit bought with my first earnings as a businessman. My enthusiasm waned as I approached the ramshackle building that housed Kazakhstan's new Ministry of Defense, recently converted from a technical college. A new country has to start with the materials it has at hand, I comforted myself. As I walked inside I saw myself as a pioneer, a founder of a new government ministry.

A young senior lieutenant, a Kazakh, met me at the entrance and told me to go upstairs to the deputy minister's office.

"You can't miss it," he said.

The informality was a relief. Safrygin greeted me warmly.

"We're honored that you could come to our tiny fortress," he said.

He offered me tea, and I relaxed on the wide sofa in his office.

The discussion started well. He asked about my work with the bank and about my family. We talked about the recent changes in Kazakhstan. Then he pulled a folder from a desk drawer.

"I'd like to show you something," he said.

The paper he spread out in front of me was a draft of an agreement between Biopreparat and the Kazakh Ministry of Defense. It outlined a plan for the joint operation of the installation we had run at Stepnogorsk.

"This is very interesting," I said at last. "But what does it have to do with me? I've left Biopreparat."

"Well," said Safrygin, "we were wondering if you would be interested in going to Stepnogorsk."

"Stepnogorsk already has a director. His name is Gennady Lepyoshkin."

"Actually, we need someone to manage the whole chain," he said.

"I'm not interested," I said.

Just then a door opened at the far end of Safrygin's office and a wiry Kazakh walked in. He looked like a soldier though he wore civilian clothes. He was about sixty years old, with thick eyebrows. Safrygin stood up. I didn't.

"Colonel Alibekov," said the visitor. "You won't mind if I join you?"

"I'm not a colonel anymore. I've left the army."

The man made a dismissive gesture. "I know that," he said.

He told me he was the chief of the defense section in the Kazakh president's administration and worked closely with Defense Minister Sagadat Nurmagambetov, who had until recently been a two-star Soviet general. He didn't give his name.

I was not pleased by the turn the conversation had taken. Nor was I pleased by the realization that this man had been listening behind the door.

"We know all about you," he continued, "and we know that

you were a capable officer. That's why we've asked you to come here today."

My heart sank.

"If you join us, we'll return you to the rank of colonel and within two weeks you will become a major general. Of course such promotions can only be made in our Kazakh constitution through a presidential decree and parliamentary approval. But I can safely guarantee it will happen."

"You don't need a major general to run a biological facility," I said.

"We plan to set up a new directorate. We want you to be its commander."

"What kind of directorate?"

"A medical-biological directorate."

"What do you mean?"

"You know exactly what we mean."

I stood up.

"Look," I said. "A treaty was signed in 1972 by countries all over the world, including the Soviet Union, that prohibits research and development into biological weapons. If your president wants to have problems with the international community in the future, then this is exactly how to do it. I recommend you forget the idea."

He flushed.

"I don't think our president needs to receive recommendations from you," he said.

"Whether he does or not, I refuse to have anything to do with this."

Kalinin must have set this up. No one in the Kazakh army would have made such an offer without his approval. It was brilliant. If I accepted, he would not only be assured of keeping the Kazakh facilities under his supervision, but he would maintain control over me. I wondered if the president of Kazakhstan knew what offer was being made in his name.

"This is not what I came here to do," I said and turned toward the door.

Realizing that he had lost the argument, the Kazakh dropped his pretense of courtesy.

"Don't think you can fool us!" he shouted. "We know your type, with your pretty suit and your Marlboros! We know all about your consorting with foreigners."

He had used the classic Stalin-era phrase in Russian—"consorting with foreigners"—that had once sent thousands to prison.

"Are you threatening me?" I asked him, my hands trembling with anger and frustration.

"I'm warning you that you may have very serious problems in the future!"

I opened the door and walked out. Behind me, I could hear a protest from the startled Safrygin, but I didn't stop to listen.

When I returned to Moscow, I felt trapped. There would be no Kazakh citizenship, and no medical or scientific career, unless I accepted the role that had been picked out for me. I couldn't even be sure I would be able to continue in private business. By turning down Safrygin's offer, I had burned my bridges in both Russia and Kazakhstan. I no longer tried to hide my intention to get as far away from Moscow as I could.

It was Savva Yermoshin who finally showed me what I had to do.

I met my old KGB friend in a hallway of the Ministry of Medical Industry in central Moscow. It was a chance encounter. I had gone there to attend a meeting of the Russian Biological Society, a scientific group whose activities I continued to participate in.

Savva seemed glad to see me. He asked how I'd been and how my family was. We hadn't seen each other since I left the agency. After some idle chat, he punched me playfully on the shoulder.

"You know, Kan, some people are nervous about you."

"Why is that?" I said, trying to keep my tone light.

"It's not really important. I keep telling them they have nothing to worry about. I tell them it's true that Kanatjan travels a lot, but he would never live in another country without his family—and of course he would never get permission to leave with them."

I said nothing.

Yermoshin laughed again. "So, are you a millionaire yet?"

"When I become one, I'll let you know," I said, matching his breezy manner.

We shook hands and I walked away. It couldn't have been easy for him to deliver that message. I had always understood that our friendship would never be placed ahead of his job.

Yermoshin suffered for that friendship. When I left he was transferred to a post outside Moscow and forced to leave the KGB. But his career wasn't hurt in the long run. He became a general in the federal tax police in a large Russian city where, I've heard, he has become a very rich man.

I owed him a debt of gratitude. I had been wondering whether to apply for passports for Lena and the children as a first step in settling abroad. Yermoshin made it clear I would never get them. The only way I could leave Russia with my family was to sneak out, like a criminal.

I thought I knew how it could be done. Over the previous months, I'd become friendly with a Russian businesswoman who lived in New York. She often traveled back and forth to Moscow, and we'd occasionally discussed business prospects in the United States. A few weeks after my return from Kazakhstan, I ran into her again at a private gathering.

I pulled her aside and quietly asked if she could do me a favor when she returned to America. I drew from my wallet the business card and telephone number given to me in December by Lisa Bronson, the U.S. Department of Defense official who had accompanied us on our trip.

I asked her to call the number from New York and to find out whether Bronson would be willing to help me emigrate to America. I had not forgotten the conversation we'd held outside the White House in December. I hoped she hadn't either.

My friend looked surprised and slightly uneasy. But she had an adventurous spirit.

"I'm planning to be in Malta in July on business," I told her when she agreed to help. "I'll call you when I get there."

I left for Malta a few weeks later. As soon as I checked into my hotel room, I picked up the phone and called New York.

I was greeted warmly.

"I spoke to your friends," she said. "They are very interested, and they say you would be welcome in the U.S."

"Thank you," I said. "Please tell them I'm coming to New York for a business trip in September. I'll call you then."

I knew my U.S. hosts would expect to hear everything I could tell them about the Soviet program in return for their help. Some of my colleagues might consider this a betrayal. But I had come to believe that my real betrayal was to have pursued a career that violated the oath I had taken as a doctor.

Back in Moscow, I told Lena what I had done. She agreed without hesitation that we should go. She was angry about my treatment in Kazakhstan and feared for my safety in Moscow.

In September I arrived in New York with my friend Naum. We had planned to spend a week in the city talking to Russian émigrés about trade possibilities, and we'd booked a double room in a hotel at Thirtieth Street and Broadway.

As soon as we arrived I called my friend. She offered to meet me anywhere, but I decided the hotel was the best place. I was nervous about venturing alone into the streets of a strange city where I couldn't speak the language and where I knew the KGB had agents. I pulled Naum aside.

"I have a favor to ask you," I said.

"Go ahead. Ask."

"I'm not sure how to put this, but there's a friend here in New York I need to see," I said awkwardly. "She's kind of an old girlfriend, and I thought we could get together for old time's sake, you know, so if you wouldn't mind staying clear of our room this afternoon . . ."

Naum winked.

"Of course. Anything to oblige."

My friend arrived a few hours later. She was nervous and spoke quickly. Lisa Bronson had given her the names of some people in Washington to call.

"They're ready whenever you are," she said. "They'll set you up so you can get paid as a consultant in biological defense. But there's one thing."

"What?"

"They want you to do it now. They think that if you go back, there's a risk you'll never be allowed to leave. They can arrange to get your wife and children out later."

I told her this was impossible. She smiled faintly.

"They thought you might not agree, but they felt they had to ask."

She gave me precise instructions for the arrangements I would have to make. The instructions involved officials in Kazakhstan, Russia, and other countries, all of whom would be compromised if their efforts could be traced. The details of my escape to America are the only secrets I have resolved to keep.

A week later I was back in Moscow. On the night of my return, I asked Lena to join me for a walk and explained the plan. I could take no risk of being overheard by KGB bugs. We decided to tell Mira but not the boys. Mira was fifteen, old enough to keep a secret, but Alan was twelve and Timur barely seven. The boys wouldn't have been able to resist bragging to their friends about a trip to America.

Discreetly, we began to prepare for our departure. I sold a few books and keepsakes, but decided to leave most of our furniture in the apartment to avoid arousing our neighbors' suspicions. I arranged with a relative for the sale of our household goods after we left. The money would be used to pay off our debts. I wanted no one to say I had left Russia to avoid creditors.

In the final weeks of September, we tried to lead as normal a life as possible. We told the boys that we would soon be going on holiday to Almaty.

The day before we were supposed to leave, we got a call from the KGB. The caller introduced himself as Captain Zaitsev of the Moscow Region KGB. He spoke in a low and pleasant voice.

"We'd like to have a talk with you," he said. "Would you mind coming down to our office?"

"I've got no time today," I said.

"How about tomorrow?"

"Ah," I said. "I'm flying to Almaty tomorrow."

"It really is urgent."

"Couldn't it wait until my return?"

"When is that?"

"Two weeks or so," I said.

He sounded unsure.

"Could I call you tomorrow, anyway?"

"Go ahead," I said.

I didn't wait for his call. The following day, we flew to Kazakhstan.

As I walked into the old apartment on Communist Prospekt I wondered if I would ever see it again.

My father had gone completely deaf, so I had to write my plans out on a pad of paper and show it to him. The old soldier read carefully. We stared at each other for a few moments, then he reached over and took my hand. He said nothing, but I understood I had his approval.

Later, I sat in the kitchen with my mother and brother. We spoke in Kazakh and Russian.

My mother asked me why I was leaving.

I told her about the surveillance, the phone taps, the great difficulty of finding the kind of work I wanted to do. And I told her about my encounter with the Kazakh Defense Ministry. Her voice was firm when she finally answered.

"You have no choice for yourself, or for your family. You have my blessing."

My experiences had touched a chord. As my brother and I listened in awestruck silence, my mother began telling us a story about the family we had never heard before. She had been a girl of ten when her father, my grandfather, was arrested by the security police on a trumped-up political charge. In prison, he contracted a fatal illness. My grandmother was allowed into the prison hospital with her two children—my mother and my uncle—to pay him a last visit.

It was a difficult moment. My grandparents had had a marriage of opposites: he was a staunch Communist and she was a member of the old Kazakh nobility, a descendant of Teuke Khan, who had unified the country in the seventeenth century and created its first legal code. My grandmother, who used to take me as a child to the mosque to imbue me with the religion of my forebears, had never been reconciled to the Socialist regime—and it was now about to kill her husband.

"He looked over at your uncle and me, and then he looked at my mother, and he told her to take us to an orphanage," she said, biting her lip. "My mother started to cry, and then I cried also because I thought that was a terrible thing to say.

"My mother asked why and he said it was the only way to save our lives. Otherwise they would soon come to arrest her too.

"But your grandmother didn't follow his advice. She took us home and hid us for months. And every night she would hear the cars coming down the street to take more people away. Each time she heard a car, she would say, There goes another person who killed your father."

My mother had tears in her eyes.

"You always have to do what you think is right."

The next evening we flew from Almaty to Moscow, where we were to board another plane and fly out of Russia. We landed a few hours before midnight. The connecting flight was not until the next morning, which presented me with a difficult decision.

Flights from Almaty arrived at Domodedovo Airport, south of Moscow. Our next flight left from Sheremetevo, the main international terminal north of the capital. The drive between the two airports was nearly two hours, and it would take us through the heart of the capital. To go directly to Sheremetevo might give our plan away to the KGB, who were sure to be watching. Our apartment was in northern Moscow, close to the highway leading to Sheremetevo. It made sense to go home first and wait.

If we were lucky, we would fool the KGB into believing we had returned as promised from our vacation.

A friend picked us up at Domodedovo. It was dark and cold, and there were few cars on the highway leading into the city.

I noticed a car following us. Its headlights glared in our rear window. When we changed a lane to let it pass it changed lanes with us. But when we turned off into our street, our tail disappeared. I took a deep breath. The first part of the plan had worked.

I paced back and forth in our apartment while the rest of my family napped. I peered out the window to check if the KGB had posted someone to wait for us. Finally, before daybreak, I woke everyone up. My friend's car was outside, puffs of white smoke trailing from the exhaust pipe.

We crept downstairs, hoping not to wake our neighbors. I held the doors open as Lena and the children bundled into the car and looked up and down the street. I couldn't see anyone.

No one followed us to Sheremetevo. My stomach continued to churn until we finally stood up in the waiting room to join the passengers boarding the plane.

It was hard to believe that we had tricked the KGB. When we settled in our seats, the stewardess's smile struck me as the most wonderful thing I had ever seen.

19

DEBRIEFING

*Russia has . . . never developed, produced, accumulated, or
stored biological weapons.*
—Address by Grigory Berdennikov, head of the Russian delegation to a November 1996
conference of signatories to the 1972 Biological Weapons Convention

One month before I came to America, Russia signed an agreement with the United States and Great Britain putting an end
to its biological weapons program. In September 1992, the three
countries agreed to work together to convert former weapons-
making facilities into centers for peaceful scientific research, to encourage scientific exchanges, and to establish procedures for
reciprocal visits to military and civilian installations. The biological arms race was on its way to becoming a closed chapter of cold
war history. Or so it seemed to the Americans who took charge of
my debriefing.

Nearly every weekday morning during my first year in the
United States, I drove to an office building in a small city in Vir-
ginia, twenty minutes' drive on Route 66 from Washington, D.C.
In a second-floor room with comfortable chairs and a large table,
I answered questions put to me by senior officials from intelligence
agencies and various branches of government, including the De-
partment of Agriculture, the State Department, the Department of
Defense, and the Arms Control and Disarmament Agency. They

generally introduced themselves, but after the fourth or fifth intro-
duction I would lose track of who they were and which agency
they came from.

At first I felt detached, but gradually I began to look forward to
the debriefings. I felt a certain relief in speaking for the first time
about the things I had kept secret for so long.

Lena remarked on the change in my demeanor. The tense government official she had lived with in Moscow was gone, replaced
by a more relaxed stranger. I would try to tell her about the day's
session after the children went to bed, but she seemed uninterested.
She wanted to forget the past.

I had expected the debriefings to be surrounded by an atmosphere of espionage and intrigue, but they were more like academic seminars. They were sometimes frustrating, especially when
it came to strategic questions, which seemed to interest my interrogators not at all.

"We're only interested in what you know," one U.S. defense an-
alyst told me, "not what you think could happen."

I understood their logic—I was an administrator and a scientist, not a military or political strategist—but their attitude seemed
to reveal a profound misunderstanding of biological weapons. My
interrogators wanted to know how much of our stockpiles and
production capacity had been shut down and which of our labs
and facilities had been destroyed. They expressed little curiosity
about the potential of the weapons we had made. Few asked me
about the specific capabilities of our anthrax, tularemia, and
plague weapons or paid more than cursory attention to our genetic
work. The emphasis on our shrinking arsenal made it clear to me
that Americans believed Russia's biological weaponry no longer
constituted a significant threat.

Slowly and reluctantly, I came to believe they were wrong.

In early 1994 I came across an article published the previous year
by Sergei Netyosov, deputy scientific director of the Vector complex. He reported that a team of scientists had successfully inserted
foreign genetic material into vaccinia, a nonpathogenic virus related to smallpox. My heart sank. This experiment was part of a

secret plan I'd authorized five years earlier to create a powerful new smallpox weapon.

I first met Netyosov in February 1989. A promising virologist in his early thirties, he was introduced to me by Lev Sandakchiev during one of my inspection trips to Siberia.

"Netyosov is one of our best people," Sandakchiev had boasted as I shook the young scientist's hand. "I'm recommending him for a promotion."

Netyosov, who held a Ph.D. in virology, belonged to an impressive new generation of civilian scientists recruited by Biopreparat in the 1980s. Sandakchiev told me he was on the verge of a breakthrough that would have as large an impact on our weapons program as the genetic experiments performed with bacteria and toxins at Obolensk.

"We believe we can create a chimera virus," he said, elliptically.

A chimera is an imaginary monster with the head of a lion, the body of a goat, and a serpent's tail. Biologists use the word to describe an organ composed of tissues of diverse genetic material. I'd never heard it applied to viral organisms before.

Netyosov's work was inspired by Western research. He had read accounts in foreign journals of a successful experiment in which scientists had inserted the gene of Venezuelan equine encephalitis (VEE), a virus that attacks the brain, into vaccinia. The experiment was part of continuing research into the viral genome, the collection of genes that code the peculiarities of every living organism, and it had significant medical implications. Understanding the genetic differences between closely related strains of viruses could help explain why some strains caused disease and others didn't. Researchers also believed that vaccines capable of immunizing people against several diseases at once could be produced by introducing the genes of one virus into another. An altered vaccinia virus, for example, could reproduce VEE cells as well as its own. The research required months, sometimes years, of painstaking work. A host virus will reject alien genes until lab technicians find a compatible place in the genome to introduce the new material.

Vaccinia's genetic structure was almost identical to the smallpox virus. If VEE could be combined with vaccinia, Netyosov observed, perhaps it could also be joined to *Variola major*, creating a "double agent," a superweapon capable of triggering both diseases at once.

Persuaded by Sandakchiev of the project's importance, I granted him permission to promote Netyosov from lab chief to deputy scientific director of the facility. Back in Moscow, I authorized a special grant of one hundred thousand rubles for the Chimera project.

The techniques used to manipulate viral genes are more complicated than those for bacteria. Some viruses, like Venezuelan equine encephalitis, are made of RNA, or ribonucleic acid, an inverted version of ordinary DNA. The gene sequences of RNA viruses must be transposed before genetic experiments can be performed. Once this has been done, the viral genome is sliced with special enzymes called restrictases and knit together with the foreign genes to create what is called recombinant DNA.

Within six months, in the spring of 1990, Netyosov reported that he had successfully inserted a DNA copy of VEE into vaccinia. Space had been found for the transplanted material in a gene of vaccinia called thymidine kinase, and it multiplied along with its new host. Netyosov's team immediately began similar genetic manipulations with *Variola major.*

At the time, I was not confident of their success. Western geneticists had discovered that when VEE and vaccinia were combined, the vaccinia appeared to lose its virulence. This was a problem for us: we did not want to weaken our smallpox weapon.

By 1990, as my attention was drawn to preparations for the foreign inspectors, I lost track of Netyosov's work. But the research continued.

Two years later, in 1996, the same team published an article in *Molecular Biology,* a journal published by the Russian Academy of Sciences. The scientists reported that they had found a space in the vaccinia genome where foreign genetic material could be inserted without affecting virulence. They claimed the purpose of this research was entirely peaceful—to explore different properties of the vaccinia virus. But what medical reason could there be for experiments aimed at preserving its virulence?

The Vector scientists had used a gene for beta-endorphin, a regulatory peptide, in their experiments. Beta-endorphin, capable in large amounts of producing psychological and neurological disorders and of suppressing certain immunological reactions, was one of the ingredients of the Bonfire program. It was synthesized by the Soviet Academy of Sciences.

In 1997, the same team reported in the Russian publication *Questions of Virology* that they had successfully inserted a gene for Ebola into the genome of vaccinia. Once again, a benign scientific explanation was put forward: they said it was an important step toward creating an Ebola vaccine. But we had always intended vaccinia to be our surrogate for further smallpox weapons research. There was no doubt in my mind that Vector was following our original plan.

One of our goals had been to study the feasibility of a smallpox-Ebola weapon.

Vector has been the official repository for Russia's smallpox stocks since they were moved from the Ivanovsky Institute in Moscow in 1994. Sandakchiev and I first tried to transfer the strains from Ivanovsky to Vector in 1990, hoping that these "legal stocks" would serve to cover up Vector's smallpox work. The Ministry of Health turned us down at the time, but four years later the Russian parliament approved the same plan with no public explanation. The transfer aroused little international attention.

The research at Vector was by no means an isolated case. In 1997 scientists at Obolensk reported in the British scientific journal *Vaccine* that they had developed a genetically altered strain of *Bacillus anthracis* capable of resisting anthrax vaccines. In earlier articles, they claimed to have developed a multi-drug-resistant strain of glanders. Both projects were initiated in the 1980s.

My American interlocutors were skeptical of my concerns. Some doubted a combined weapon was possible. Scientists whom I respect wondered why anyone would want to make such a weapon.

Smallpox and Ebola, they pointed out, were each sufficiently lethal on their own. Dr. Peter Jahrling of USAMRIID, who was present at some of my early debriefing sessions, has called the concept "sheer fantasy."

I have no way of knowing whether a combined Ebola-smallpox agent has been created, but it is clear that the technology to produce such a weapon now exists. To argue that these weapons won't be developed simply because existing armaments will do a satisfactory job contradicts the history and the logic of weapons development, from the invention of machine guns to the hydrogen bomb.

I told my debriefers that Russia's biological labs should be as carefully monitored as its nuclear arsenal. I was told in turn that it is wrong to conclude intentions from the nature of scientific research, and that the work being conducted in Russia should be accepted as peaceful until there is a compelling reason to think otherwise.

Throughout my career, I had worried that American scientists would surpass us. Now I found myself struggling to persuade them how far the science of germ warfare had come. It wasn't until Bill Patrick walked through the door two months after my first debriefing that I felt someone understood what I was trying to say.

Patrick handed me his business card as soon as we were introduced. I couldn't read a word, but when I saw the skull and crossbones over his name, I started to laugh. The card, I later found out, identified his occupation with a single word: "bioweaponeer."

Patrick, then in his late sixties, had retired from Fort Detrick, where he had made a smooth transition from supervising the U.S. Army's biological warfare "product development" division to formulating methods for the protection of soldiers from the weapons he and his associates had made. He had become a consultant on biodefense, participating in the first United Nations team of arms monitors sent to Iraq in 1992. The difference in our ages and backgrounds evaporated as we shared the secrets of our former profession. We had tackled many of the same scientific problems. When I gave him details of the recipes for our weapons, he buried his head in his hands.

Patrick knew as well as I did that improvements in the cultivation, concentration, and delivery of biological agents since the closure of the U.S. program presented Americans with a grave security risk.

Despite the Kremlin's pledge, Russian military commanders neither opened their biological facilities to foreign inspection nor disavowed their commitment to biological warfare.

"We are restoring what was destroyed between 1986 and 1989," declared Major General Anatoly Khorechko, who now runs Compound 19 in Yekaterinburg (Sverdlovsk), in an interview in 1997 with the base's internal newspaper. His remark was reprinted in *Top Secret,* one of Russia's best-informed investigative journals, as part of a lengthy report on the facility. The article noted that Compound 19 had also purchased reactors and other pharmaceutical equipment from Japan.

Signals have come from other facilities. The vice governor of the Penza region declared in 1997 that his area "will soon have biological weapons."

I am convinced that a large portion of the Soviet Union's offensive program remains viable despite Yeltsin's ban on research and testing. Assembly lines were destroyed at Omutninsk, Berdsk, Stepnogorsk, Kurgan, and Penza as a result of Gorbachev's decree. These facilities were transformed into pharmaceutical and pesticide plants, but only a few alterations would be required for them to serve as weapons-assembly lines again. In some cases, it would only take a few months. Stepnogorsk is the only facility at which weapons production has been foreclosed. In 1998, the Kazakhstan government agreed to dismantle the entire facility in exchange for millions of dollars from the United States as part of a broader initiative to dismantle the old Soviet nuclear and biological weapons complexes.

Vector, Obolensk, and the Institute of Ultra-Pure Biopreparations in Leningrad remain under state control. Equipment design and manufacturing plants such as the Precision Machinery Bureau outside Leningrad, the Bureau of Instrument Controls and Automation at Yoshkar-Ola, and the branches of the giant Biomash conglomerate have been retooled for civilian work. Some of these have biodefense contracts with the army.

Offensive research at institutes run by the Academy of Sciences and Ministries of Health and Agriculture has ended, and our stockpiles of plague, tularemia, and smallpox have been destroyed. Nonetheless, there is persistent evidence that Russia continues to place a high value on its old biological warfare infrastructure.

The commanders of the three principal military biological facilities—in Yekaterinburg (Sverdlovsk), Sergiyev Posad (Zagorsk), and Kirov—were promoted from colonel to general between 1992 and 1994. The Yeltsin government claimed this was a recognition of the importance of biodefense work, but if the plants are only making vaccines, why are they sealed off from the public? The U.S. has one comparable military facility—USAMRIID—and it regularly grants permission for visits.

Many former commanders and bureaucrats in the Soviet biological war machine continue to hold important government positions. General Valentin Yevstigneyev, who led the Fifteenth Directorate during my last year in Biopreparat, is deputy director of the Russian army's Nuclear, Biological and Chemical Arms Control Directorate. Hard-liners who were once passionate advocates of biological weapons are regaining influence in Moscow, among them Yury Maslyukov, the former military-industrial chief, now deputy prime minister of Russia. Yeltsin's successors may be less inclined to accept Western curbs on the country's military potential.

Not long ago an American official who had just returned from a visit to Moscow showed me a brochure celebrating the twenty-fifth anniversary of the founding of Biopreparat. Prominently featured on the brochure was a photo of Yury Kalinin, who was also celebrating his sixtieth birthday. To my surprise, I learned that Kalinin was still a general, five years past the normal retirement age for Russian officers. How, I wondered, could Russia maintain that Biopreparat was solely devoted to peaceful research when its director continued to hold a military rank?

As if to emphasize the point, my old boss recently decided to send me a message.

On a muggy Friday evening last August, a man in a charcoal gray suit walked into the lobby bar of the Ritz-Carlton Hotel in Pentagon City, a few miles from where I now live. Hesitating at the doorway, he peered at the crowd like a passenger dropped off in a strange port.

I felt a twinge of nervousness as he threaded his way toward my table. He was the director of a Biopreparat research facility—the first person from my former circle to cross my path in five years.

A State Department acquaintance had told me he was visiting Washington to raise funds for his institute. On an impulse, I decided to call him at his hotel and suggested we meet for a drink. He was reluctant, but he called back several hours later and agreed to meet me at the Ritz-Carlton bar.

I was told by friends in Moscow that the KGB opened an intensive investigation immediately after I left. The United States did not publicize my defection, choosing instead to comply with the secrecy demanded by Moscow under the terms of the trilateral agreement negotiated after Pasechnik's defection. There were reports in 1993 and 1994 about a "second biological defector," but my identity was never revealed. Nevertheless, the breadth of the KGB investigation suggested Moscow was preparing a dossier to discredit me, should it ever become necessary. Nearly everyone I had known or worked with in my entire career at Biopreparat was interrogated, and some of my colleagues had suffered from their association with me.

The jazz band was beginning a new set when we shook hands. My friend looked at me with what I took to be amusement. I was wearing the summertime uniform of suburban America: a sports shirt and casual slacks. He was in a dark, ill-fitting suit, too heavy for the heat.

"So," he said, glancing at the faces nearby as he took a seat. "Which ones are yours and which are ours?"

I laughed. It was a line whose black humor only two ex–Soviet bureaucrats could appreciate. But he'd drawn a line between us: I was now one of "them."

I had a glass of wine. He ordered a martini and we settled into what I hoped would be a conversation about old times. It wasn't until I asked about his current projects that he grew animated. He began to tell me about a "biological defense project" funded by the

Ministry of Defense. I started to talk about about my own work when he put his hand on my arm.

"You don't have to explain yourself," he said. "I know why you came to America. You've made your decision, and I've got no problems with that. I'm not someone who thinks you're a traitor."

He let the statement hang in the air, as if to remind me that others did. Then he gave a dismissive shrug and tried to smile.

"Kan," he said, "I hope you don't mind if I inform Kalinin that we have spoken?"

I couldn't hide my astonishment. Until that moment, I'd assumed that Kalinin had retired. My friend's initial hesitation about seeing me, followed by his decision to come to the hotel, now seemed ominous. Had he requested permission to meet with the "traitor"? There would have been enough time for him to call Kalinin before coming to the Ritz-Carlton.

"Of course I don't mind," I said uneasily. "How is the general anyway? I thought he would have left Biopreparat by now."

My friend shook his head.

"He's the same."

We fell into an awkward silence.

"You know," I said at last, "I'd love to go back some day, maybe after I get my U.S. citizenship."

"That wouldn't be a good idea," he said at once.

"Why not?"

He stared at his glass.

"Kalinin has been telling people that if you ever return to Moscow you won't be leaving," he said.

"What is that supposed to mean?"

"He says you betrayed our secrets."

"So he'll have me arrested?"

"Worse."

I was beginning to regret the whole encounter.

"What can he do?"

My friend concentrated on his martini.

"It would be no problem to find someone to kill you," he said.

"This is ridiculous."

"It's not ridiculous," he said stubbornly. "You don't know

what it's like in Moscow these days. You can get someone killed for ten thousand dollars."

My dubious look only served to increase his agitation. He took out his handkerchief and wiped his face.

"Okay," I said finally. "Thanks for the advice."

He rose from the table, saying he had an early flight the next morning. I stood up to shake his hand. We promised to stay in touch and I watched with relief tinged with irritation as he disappeared into the crowd.

I wondered if I would ever be free of my past. The idea of Kalinin contracting a mafia hit man for my murder seemed ludicrous. Five years was a long time to nurse a grudge. Why would anyone in Moscow care about my knowledge of a program that supposedly no longer existed?

Then it came to me. My old colleagues were not worried about what I could tell Americans about the past; they feared my knowledge of the present.

Kalinin is not the only Russian aggravated by the role I've played since coming to America. Oleg Ignatiev, the former chief of biological warfare at the Military-Industrial Commission and now member of a Russian presidential committee on arms control, told one of his American guests that he had bought two pet monkeys.

"I've named one of them Pasechnik and the other Alibekov," he said, "and when I'm in a bad mood, I beat one or the other."

My formal debriefing sessions were over by the end of 1993. I continued to meet with senior officials who asked to see me from time to time, and gradually my concerns about Russia gained greater acceptance in the intelligence and defense communities. Yet even those who shared my doubts that Russia had completely abandoned biological warfare research believed the risk of a revived program remained small.

They argued that Moscow placed too much value on its burgeoning partnership with the United States to risk alienating Washington. Besides, they added, there was no reason for the Kremlin to waste its scarce resources on biological weaponry when the only threats Moscow faced from Europe and the United States were the insistent demands of creditors. My response was that some of Moscow's principal security concerns today could best be addressed with biological weapons.

Russia's army is demoralized. The disastrous war in Chechnya exposed the shortcomings of conscript troops, and officers have gone without pay for months at a time. Yet the weakened Russian military machine confronts a greater variety of challenges than it ever faced during the cold war. These include armed separatist movements in the Caucasus, civil wars in central Asia, the spread of Muslim fundamentalism from Iran and Afghanistan, and pressure from a resurgent China. The late-twentieth-century specter of "total war" has been replaced by the growth of ethnic, nationalist, and religious conflicts. Biological weapons can play an important part in such conflicts, often compensating for the weakness or ineffectiveness of conventional forces.

Several months before Soviet forces withdrew from Afghanistan in 1989, I was told by a senior officer in the Fifteenth Directorate that the Soviet Union used biological weapons during its protracted struggle with the mujaheddin. He said that at least one attack with glanders took place between 1982 and 1984, and there may have been others. The attack, he claimed, was launched by Ilyushin-28 planes based in military airfields in southern Russia.

It was a casual remark, but the officer was evidently proud of the operation, and of the fact that he could tell me a secret about a project I knew nothing about.

When I mentioned this conversation during one of my debriefing sessions, an American intelligence official in the room was visibly startled. She told me there had been periodic reports of disease outbreaks among guerrilla groups in Afghanistan during the war. No one had ever come up with an explanation.

I grew more convinced after reading an April 1998 article in *Top Secret* that disclosed that the army facility in Sverdlovsk had manufactured "anti-machinery" biological weapons in the 1980s for use in Afghanistan. I knew of no projects involving such agents when I became deputy director of Biopreparat, but one of the bacterial strains investigated in the 1970s for its corrosive properties came from a bacterial genus known as *Pseudomonas*. The source for *Top Secret*'s report could have unwittingly, or intentionally, confused it with glanders, which was then classified by biologists as part of the same genus. Although it has subsequently been given a different scientific name, glanders was known at the time as *Pseudomonas mallei.* The pathogen is not usually lethal to humans, but we considered it an excellent battlefield weapon. Sprayed from a single airplane flying over enemy lines, it could immobilize an entire division or incapacitate guerrilla forces hiding in rugged terrain otherwise inaccessible to regular army troops—precisely the kind of terrain our soldiers faced in Afghanistan.

I was cautioned by government officials against speaking out too bluntly against Russia. Even if I was right, they argued, there was no point in pushing Moscow further than it was willing or able to go.

"Perhaps there are questionable activities going on," one conceded, "but for the moment, diplomacy requires us to keep silent."

20

BUYERS AND SELLERS

In the summer of 1995, I received a call from a man who introduced himself as a representative of the government of South Korea. Explaining that he had been given my name by a mutual friend, he said he needed my help urgently. We met at a crowded open-air café in Bethesda, Maryland.

He was courteous and friendly and came to the point at once.

"Your knowledge is extremely valuable to us," he said. "You could make a lot of money telling us what we need to know. We would like to invite you to Seoul."

When I asked him what "knowledge" he was interested in, he told me his government had evidence of a biological warfare program under way in North Korea, which has been trying to destroy its southern neighbor for the past four decades.

"We've had to worry about their army and their nuclear weapons and their saboteurs," he said. "Now we need to learn how to defend ourselves against this biological threat. You can be certain your help will be well rewarded. The South Korean minister of defense is a close personal friend."

I suggested he apply for my services through official channels in Washington, pointing out that I was still under some obligation to those who had helped me escape from Russia. He shrugged this off. Seoul and Washington were close allies, he said. No one would mind. I insisted, and never heard from him again.

South Korea was not the only country to ask for my help. I was approached by a member of the French embassy after delivering a lecture in Boston in mid-1998. He invited me to lunch with embassy officials in Washington to discuss "biodefense issues." I told him it was a sensitive subject and asked him to send a formal letter to the research and development company where I now work. No letter appeared. A similar request came from a friend with connections to the government of Israel.

Growing fears of a biological attack by a hostile neighbor or a terrorist group have created a booming mini-industry of biodefense consultants. Biodefense requires knowledge of the capabilities of pathogenic agents, their means of delivery, and potential effects. This knowledge is also the key to developing offensive weapons. I evaded these requests in part because I didn't want to turn into an unwitting conduit for new bioweapons work. Fortunately, I had an alternative—a job I liked—and my family was well taken care of. But the monetary rewards for cooperation would have been high.

The services of an ex-Biopreparat scientist would be a bargain at any price. The information he could provide would save months, perhaps years, of costly scientific research for any nation interested in developing, or improving, a biological warfare program. It is impossible to know how many Russians have been recruited abroad, but there is no doubt that their expertise has been attracting bidders. At least twenty-five former specialists in the Soviet Union's biological warfare program are now in the United States. Many more have gone to Europe and Asia or have simply dropped out of sight. I've heard that several went to Iraq and North Korea. A former colleague, now the director of a Biopreparat institute, told me that five of our scientists are in Iran. *The New York Times* reported in December 1998 that the Iranian government dispatched a "scientific advisor" attached to the office of the presidency to Moscow to recruit former scientists from our program. In May 1997, more than one hundred scientists from Russian laboratories, including Vector and Obolensk, attended a Biotechnology Trade Fair in Tehran. Sandakchiev told me soon after that Iranians had visited Vector a number of times and were actively promoting scientific exchanges. Last year, *Top Secret* reported that a Biopreparat official turned up at the Chinese embassy in Moscow to offer his services.

The disastrous economic conditions in Russia have driven many of our brightest scientists and technicians to seek work wherever they can get it. In some labs, scientists haven't been paid for months. I know of one leading researcher who sold flowers on the Arbat Mall in Moscow to feed his family.

The West is worried, with good reason, about lax security at Soviet nuclear installations. The vulnerability of our biological arsenal should also raise concern. A vial of freeze-dried powder takes up less space than a pack of cigarettes and is easy to smuggle past an inattentive guard. It happened when I was at Biopreparat, when security was at its peak. Biological agents once kept secure in government facilities are rumored to be circulating freely in the Russian criminal underworld.

Every agent developed in our labs came with a detailed set of instructions outlining the production process from seed culture to drying and assembly. The complete recipe for our anthrax weapon filled twelve volumes. To save storage space, the Fifteenth Directorate decreed in 1991 that all final formulations be microfilmed and sent to military facilities at Sergiyev Posad, Kirov, and Yekaterinberg. Those facilities are closely guarded, but a military scientist in desperate economic straits might find it hard to resist the temptation to smuggle out a tiny roll of microfilm.

The Kremlin has revived travel restrictions for those considered privy to state secrets, but our scientists don't always have to leave home to find a market for their talents. Not long ago, I obtained a copy of an advertising flyer printed by a Moscow-based company called Bioeffekt Ltd. It offered, by mail order, three genetically engineered strains of tularemia. According to Nikolai Kislichkin, identified in the flyer as company president, the strains contained genes responsible for increasing the virulence of tularemia and melioidosis. Boasting that they were produced by "technology unknown outside Russia," Kislichkin said they would be useful for the creation of vaccines. He was well aware that they could also be used for less benign purposes. Kislichkin had been a scientist at Obolensk.

Dozens of small privately owned pharmaceutical companies like Bioeffekt have flourished in Russia since the Soviet collapse. They represent another channel through which the techniques, the knowledge, and even the strains we developed have spread beyond the borders of the old Soviet Union, contributing to an alarming proliferation of biological weapons since the end of the cold war.

When I became deputy director of Biopreparat, secret reports on the global state of biological weapons research were sent to me twice a month. They were prepared by a number of intelligence agencies, including the KGB, the GRU, and Medstatistika, a covert research institute at the Ministry of Health.

To our knowledge, none of our satellites in Eastern Europe ran biological weapons programs, though some of our fermenting and drying equipment was manufactured in East Germany. Espionage reports provided evidence of a biowarfare program in Iraq as of 1988 and identified a large biological warfare research complex near Pyongyang, the capital of North Korea. In northwestern China, satellite photos detected what appeared to be a large fermenting plant and a biocontainment lab close to a nuclear testing ground. Intelligence sources found evidence of two epidemics of hemorrhagic fever in this area in the late 1980s, where these diseases were previously unknown. Our analysts concluded that they were caused by an accident in a lab where Chinese scientists were weaponizing viral diseases. A "BW related" facility was identified in Germany (in Münster) and two in France, but much slipped by unnoticed by our intelligence gatherers.

When Yury Ovchinnikov died in 1987, I joined a group of Biopreparat scientists at his funeral services in Moscow. The conversation eventually turned to Cuba's surprising achievements in genetic engineering. Someone mentioned that Cuban scientists had successfully altered strains of bacteria at a pharmaceutical facility just outside of Havana.

"Where did such a poor country get all of that knowledge and equipment?" I asked.

"From us, of course," he answered with a smile.

As I listened in astonishment, he told me that Castro had been taken, during a visit to the Soviet Union in February 1981, to a laboratory where *E. coli* bacteria had been genetically altered to produce interferon, then thought a key to curing cancer and other diseases. Castro spoke so enthusiastically to Brezhnev about what he had seen that the Soviet leader magnanimously offered his help. A strain of *E. coli* containing the plasmid used to produce interferon was sent to Havana, along with equipment and working procedures. Within a few years, Cuba had one of the most sophisticated genetic engineering labs in the world—capable of the kind of advanced weapons research we were doing in our own.

General Lebedinsky visited Cuba the following year, at Castro's invitation, with a team of military scientists. He was set up in a ten-room beach-front cottage near Havana and boasted of being received like a king. An epidemic of dengue fever had broken out a few months earlier, infecting 350,000 people. Castro was convinced that this was the result of an American biological attack. He asked Lebedinsky and his scientists to study the strain of the dengue virus in special labs set up near the cottage compound. All evidence pointed to a natural outbreak—the strain was Cuban, not American—but Castro was less interested in scientific process than in political expediency.

Shortly after Lebedinsky returned to Moscow, Castro accused America of attacking Cuba with biological agents. A public outcry ensued, but evidence was unpersuasive. Lebedinsky was asked by the KGB to keep his work to himself. This was not the first time Castro had made such a claim; nor was it the last. Cuba has accused the United States twelve times since 1962 of staging biological attacks on Cuban soil with antilivestock and anticrop agents. The latest claim, filed with the United Nations in 1997, was the first ever submitted to the United Nations under Article 5 of the Biological Weapons Convention. It accused the United States of disseminating *Thrips palmi,* a plant-destroying insect, with crop spraying planes. The United States countered that the planes were ferrying ordinary pesticides to coffee plantations in Colombia.

Kalinin was invited to Cuba in 1990 to discuss the creation of a new biotechnology plant ostensibly devoted to single-cell protein. He returned convinced that Cuba had an active biological weapons program.

The situation in Cuba illustrates the slippery interrelation between Soviet support of scientific programs among our allies and their ability to develop biological weapons. We spent decades building institutes and training scientists in India, Iraq, and Iran. For many years, the Soviet Union organized courses in genetic engineering and molecular biology for scientists from Eastern Europe, Cuba, Libya, India, Iran, and Iraq, among others. Some forty foreign scientists were trained annually. Many of them now head biotechnology programs in their own countries. Some have recruited the services of their former classmates.

In July 1995, Russia opened negotiations with Iraq for the sale of large industrial fermentation vessels and related equipment. The model was one we had used to develop and manufacture bacterial biological weapons. Like Cuba, the Iraquis maintained the vessels were intended to grow single-cell protein for cattle feed. What made the deal particularly suspicious was an additional request for exhaust filtration equipment capable of achieving 99.99 percent air purity—a level we used only in our weapons labs.

Negotiations were called off by the time reports of the deal surfaced in the Western press, but a United Nations employee told me Iraq obtained the equipment it needed elsewhere. United Nations Special Commission inspection teams, established after the Gulf War to monitor the dismantlement of Iraq's chemical and biological weapons program, had not been able to find this equipment by the time they were ejected from Iraq in late 1998. Many similar deals have gone ahead undetected.

One of the Russian officials involved in negotiations with Iraq was Vilen Matveyev, formerly of the Fifteenth Directorate and later a senior deputy at Biopreparat. Matveyev specialized in developing weapons-manufacturing equipment. He is still working as a technical adviser to the Russian government.

In 1997 Russia was reported to be negotiating a lucrative deal with Iran for the sale of cultivation equipment including fermenters, reactors, and air purifying machinery. The equipment was similar to that which was offered to Iraq.

I have tried in this book to show how the Soviet Union developed a sophisticated biological warfare program and hid it from the world, but the extent of our achievement shouldn't lead anyone to assume that biological warfare is beyond the grasp of poorer nations.

In 1989, I visited New Delhi with a large Soviet delegation to conclude an agreement on the exchange of pharmaceutical equipment. The atmosphere had been cordial on both sides, reflecting the deepening alliance between Mikhail Gorbachev and India's leader, Rajiv Gandhi. Scientific exchanges with India were not uncommon. As early as the 1960s, Lev Telegin, who later became First Deputy Minister of Medical Industry, oversaw a project to build a huge production plant for vaccines and antibiotics four hours by car outside of Ahmadabad. The Soviet Union had been supporting India both militarily and scientifically ever since.

Negotiations took place at the State Department of Biotechnology, an agency responsible for coordinating the research and production of vaccines, not far from the main government complex. One of the two administrators was a military officer who came to Vector on an official visit the following year. Heavily armed soldiers were stationed inside the facility. As we were shown through the building, I noticed several sections were closed off with coded locks.

I rose to go to the bathroom and was followed by one of the plainclothes guards sitting behind us at the conference table. I could hear his footsteps echoing in the corridor. As I reached the bathroom, he followed me inside. I found it hard to understand why an official of a friendly nation couldn't be trusted to go to the bathroom on his own. I was outraged at first, but eventually I calmed down. We went to similar lengths, after all, to protect our facilities from outsiders.

My colleagues agreed that the unusually tight security and ubiquitous military presence suggested biowarfare activity. From then on, I paid closer attention to the facilities we were shown.

On a subsequent trip to complete our negotiations, we were

taken to a small biological complex at Mukteswar, a remote village in the Himalayan Mountains close to the border with Nepal. Security there was even tighter than in New Delhi. We were asked not to enter any of the buildings on our own. One member of our delegation asked why.

"It's too dangerous," we were told. "We're studying viruses there. Besides, most of the equipment is old. There's nothing interesting to see."

Nations engaged in chemical or nuclear weapons programs almost invariably add biological weapons to their inventory. This is particularly true in cases when a country is bent on doing everything possible to protect itself against its neighbors. India faces two hostile neighbors on its borders—China and Pakistan—with whom it has fought repeatedly over the last fifty years. Its decision to conduct nuclear tests in May 1988 showed it was willing to defy international opinion for the sake of national security.

A report submitted by the U.S. Office of Technological Assessment to hearings at the Senate Permanent Subcommittee on Investigations in late 1995 identified seventeen countries believed to possess biological weapons—Libya, North Korea, South Korea, Iraq, Taiwan, Syria, Israel, Iran, China, Egypt, Vietnam, Laos, Cuba, Bulgaria, India, South Africa, and Russia. More have joined the list since.

Ordinary intelligence and surveillance techniques cannot prove the existence of a biological warfare program. Even the highest resolution satellite imagery can't distinguish between a large pharmaceutical plant and a weapons complex. The only conclusive evidence comes from firsthand information. Western suspicions about the Soviet program were only confirmed with Pasechnik's defection. South Africa's efforts to develop biological assassination agents were first revealed when the program's director testified before the Truth and Reconciliation Commission, a government-appointed panel investigating the abuses of the apartheid era. It was not until Saddam Hussein's son-in-law, Hussein Kamel, defected in 1995 that the West came to know the extent of Iraq's germ warfare program. Kamel confirmed that Iraq had begun its program a decade earlier at the Muthanna State Establishment,

eighty miles northwest of Baghdad, where researchers were cultivating anthrax, botulinum toxin, ricin, and aflatoxin, a poison that
can be found in corn, pistachio nuts, and other crops. By the time
the United Nations inspectors identified and destroyed Iraq's principal germ warfare facility at Al Hakun in 1996, Iraq had amassed
hundreds of thousands of gallons of liquid anthrax and many
other pathogens. Iraq is still suspected of harboring germ weapons
and continues to resist all attempts to probe further.

Some Western analysts maintain that evidence of biological
warfare research is not proof that viable weapons are being produced. They argue that countries with "low-tech" scientific establishments often can't make weapons or delivery systems matching
their ambitions. But even the most primitive biological weapons
lab can produce enough of an agent to cripple a major city.

On March 20, 1995, members of the Aum Shinrikyo cult sprayed
sarin gas in the Tokyo subway. Twelve people were killed and over
fifty-five hundred were injured. Subsequent testimony at the cult
leaders' trial revealed that Aum Shinrikyo had tried nine times between 1990 and 1995 to spread botulinum toxin and anthrax in
the streets of Tokyo and Yokohama. Seiichi Endo, a onetime graduate student in genetic engineering who headed the cult's "Ministry of Health and Welfare," testified that their delivery
methods—spraying the agents from a rooftop or from the back of
a van—had proven faulty, and their strains were not sufficiently
virulent. But it is not difficult to find better strains.

Viruses and bacteria can be obtained from more than fifteen
hundred microbe banks around the world. The international scientific community depends on this network for medical research
and for the exchange of information vital to the fight against disease. There are few restrictions on the cross-border trade in
pathogens.

I was told by American biowarfare experts that Iraq obtained
some of its most lethal strains of anthrax from the American Type
Culture Collection in Rockville, Maryland, one of the world's
largest "libraries" of microorganisms. Iraqi scientists, like ours,
discovered which strains to order by reviewing American scientific
journals. For thirty-five dollars they also picked up strains of tu-

laremia and Venezuelan equine encephalitis once targeted for weaponization at Fort Detrick.

Six weeks after the Aum Shinrikyo attack, Larry Harris, a member of a white supremacist group in Ohio, ordered three vials of plague from the American Type Culture Collection catalog. Requests must be made on the letterhead of a university or laboratory, so Harris designed his own stationery. The order was being processed when he phoned less than two weeks later to ask why it was taking so long. Company officials grew suspicious—legitimate medical researchers would have known it normally takes more than a month to fill an order—and eventually turned him in.

Partly as a result of this incident, Congress passed a law in April 1996 requiring germ banks and biotech firms in the United States to check the identity of all prospective buyers. This is a useful deterrent, but it has not closed off opportunities for trade. Whether cultured by state-run organizations, terrorist groups, or crazed individuals, biological weapons have moved from a closely held secret of the cold war to the international marketplace.

On December 27, 1998, in Pomona, California, a suburb of Los Angeles, 750 people were quarantined after police received a call claiming that anthrax had been released in the Glass House nightclub. It turned out to be a hoax, but the men and women in the club were quarantined for four hours. This was the last in a series of anthrax hoaxes—more than a dozen over the previous two weeks, the last two weeks of December 1998. How much worse will things be in December 1999?

21

BIODEFENSE

BIOLOGICAL ATTACK INDICATORS:

The following section contains indicators to help identify whether a biological attack has occurred.

Indicators—Description

An unusual number of sick or dead people and animals within an area or location. Any number of symptoms can be present in a suspected biological attack. As a first responder you should consider assessing (polling) the local area hospitals to see if additional casualties with similar symptoms have been observed.

Casualties can present in minutes, hours, days, and even weeks after an incident has occurred.

The time required before symptoms are observed in a biological attack is dependent upon the actual agent used. . . . When considering biological attacks from the perspective of a first responder it is important to note that, with the exception of some toxins, any manifestations of the attack are likely to be delayed.

—From *First Responders Chem-Bio Handbook: A Practical Manual for First Responders*, 1998

We may not realize until too late that we have become the victims of a biological attack. It is not until days or weeks after such an attack has taken place—after the first wave of deaths—that we will most likely recognize its occurrence. Few terrorists will choose to warn us of their activities. A small amount of Marburg or Ebola released in the subway system of Washington, D.C., Boston, or New York, or in an airport, shopping mall, or financial center, could produce hundreds of thousands of victims.

In the past twenty years, scientists have created antibiotic-resistant strains of anthrax, plague, tularemia, and glanders. Biopreparat research proved that viruses and toxins can be genetically altered to heighten their infectiousness, paving the way for the development of pathogens capable of overcoming existing vaccines. The arsenal of a determined state or terrorist group could include weapons based on tularemia, anthrax, Q fever, epidemic typhus, smallpox, brucellosis, VEE, botulinum toxin, dengue fever, Russian spring-summer encephalitis, Lassa fever, Marburg, Ebola, Bolivian hemorrhagic fever (Machupo), and Argentinean hemorrhagic fever (Junin), to name a few of the diseases studied in our labs. It could also extend to neurological agents, based on chemical substances produced naturally in the human body.

It is easier to make a biological weapon than to create an effective system of biological defense. Based on our current level of knowledge, at least seventy different types of bacteria, viruses, rickettsiae, and fungi can be weaponized. We can reliably treat no more than 20 to 30 percent of the diseases they cause.

Few Americans are aware that they are living under a state of national emergency relating to weapons of mass destruction. On November 14, 1994, President Clinton issued Executive Order 12938, asserting that the potential use of nuclear, biological, and chemical weapons by terrorist groups or rogue states represented "an unusual and extraordinary threat to the national security, foreign policy, and economy of the United States." The order made it illegal for Americans to help any country or entity to acquire, design, produce, or stockpile chemical or biological weapons and placed the country in a state of emergency. It has been renewed every year since. In 1998, it was amended to include penalties for trafficking in equipment that could indirectly contribute to a foreign germ warfare program.

In June 1995, Clinton outlined a new policy against "superterrorism"—terrorism involving weapons of mass destruction. Today, as a result of that policy, the Departments of Defense, Energy, and State, together with the FBI and the CIA, oversee a wide network of military and civilian agencies dedicated to identifying biological or chemical attacks and to coping with their consequences. Among those agencies are USAMRIID, the Centers for Disease Control in Atlanta, the Department of Agriculture's Exotic Disease Laboratory, the Lawrence Livermore National Laboratory in California, and the Sandia National Laboratory in New Mexico. Meanwhile, existing military units such as the Marine Corps Chemical and Biological Incident Response Force (CBIRF), the army's Technical Escort Unit, and the Department of Energy's Nuclear Emergency Search Team (NEST) have been upgraded.

In 1997 the government authorized a $52.6 million Domestic Preparedness Program for emergency response teams or "first responders" in 120 selected cities across the United States. Police, fire department, and public health officials in those cities will receive special training and equipment to help them contain and combat biological and chemical terrorism. Denver was the first city chosen for the pilot program. New York, Los Angeles, Chicago, Houston, Washington, Philadelphia, San Diego, and Kansas City were added to the list in 1998 and are expected to be fully operational by the end of 1999. Parallel efforts are under way to explore methods of strengthening the security of public buildings with tamper-proof ventilation systems and improved air filtration units.

On May 22, 1998, in a speech to the graduating class at the United States Naval Academy at Annapolis, President Clinton proposed a five-year $420-million initiative to create a reserve stockpile of vaccines and antibiotics to protect Americans against biological attacks. The initiative was intended to broaden an immunization program introduced five years earlier to safeguard American troops on the battlefield. Since then, biological terrorism has become one of America's principle security concerns. In January 1999, after a year in which the American military attacked Sudan, Afghanistan, and Iraq and dozens of anthrax scares were reported throughout the country, Clinton unveiled a new plan for combatting bioterrorism at home. "The fight against terrorism is far from over," he said in a speech at the National Academy of Sciences, "and now terrorists seek new tools of destruction. The enemies of peace realize they cannot defeat us with traditional military means, so they are working on new forms of attacks."

Clinton announced new government spending of $1.4 billion in fiscal year 2000 to create and strengthen urban emergency response teams, protect government buildings, improve the nation's ability to detect and diagnose disease outbreaks linked to biological agents, and increase national stockpiles of vaccines and antibiotics. Close to $400 million would be spent on detection technology and research into new vaccines.

Donna Shalala, the secretary of health, spoke after Clinton. "This is the first time in American history in which the public health system has been integrated directly into the national security system," she said. The President warned Americans not to panic. He insisted that new intelligence needs would not infringe on civil liberties.

America has done more than any other nation to protect civilians from biological weapons. But it is not clear, for all of its efforts, that its citizens are any safer.

No one can seem to agree on the best approach to biodefense. The First Responders Program has already encountered criticism. "This approach merely displaces risk, and forces the terrorist, who is often flexible, to select a 'softer' target, in this case a city which did not receive the needed training and equipment," Frank Cilluffo, director of the Terrorism Task Force at Washington's Center for Strategic and International Studies, said in testimony before Congress on October 2, 1998. The real problem is that it assumes an identifiable scene of attack; biological weapons will most likely be deployed in secret and leave no trace.

Early biodefense exercises revealed serious flaws and a general confusion as to how to coordinate local and federal efforts. In a simulated attack staged in New York City in 1998, nearly all of the members of an emergency unit dispatched to the scene "died" because they were insufficiently protected. "They did all the right things," a federal official who watched the exercise told *The New*

York Times. "But the scenario utterly defeated them." The emergency teams were hampered by their inability to identify which biological agents had been used.

Early detection is a key element of biological defense. Depending on the agent and the manner of its dissemination, physicians and emergency rescue teams may have as little as an hour to figure out how to contain a looming medical catastrophe.

The United States has been investigating detection systems with varying degrees of success since World War II. Most methods involve the exposure of vials or petri dishes containing laboratory-grown cultures to air samples from a suspected target area. This can be a laborious process. A field monitoring device used during the Gulf War took between thirteen and twenty-four hours to make a positive identification. For botulinum toxin, one of the staples of Iraq's arsenal, this would already be too late. Technology has improved since then. The Biological Integrated Detection System (BIDS) cut the time to only thirty minutes, but it can so far only determine the presence of anthrax, plague, botulinum toxin, and staphylococcus enterotoxin B.

In September 1998, Clinton and Yeltsin agreed in Moscow on a program of "accelerated negotiations" to strengthen the Biological Weapons Convention. The United States has taken the lead in efforts to bring the treaty up to date. A so-called ad hoc group of countries met four times in 1998 to draft an amendment for mandatory inspections in countries suspected of developing or harboring biological weapons. Other measures discussed include requiring countries to open their biological facilities to regular visits from international inspectors and setting up a unit to investigate suspicious outbreaks of disease. Five more meetings of the ad hoc group are scheduled for 1999. Areas of discussion will broaden to include methods of blocking the transfer of sensitive technology on the Internet, at scientific conferences, and through student exchange programs.

The amendments, if approved, would provide a useful curb against future proliferation. But a determined state is likely to find ways to circumvent them. Consider Iraq, where the United Nations Special Commission has been given virtually unlimited authority to monitor every aspect of the disarmament program imposed by the U.N. Security Council since the Gulf War. These measures are far tougher than any contemplated under the ad hoc process and constitute an intrusion into national sovereignty that would not be tolerated by most countries. Yet despite the periodic threat (and implementation) of military strikes, Iraq has defied U.N. inspections at will. How likely are we then to impose a similar degree of compliance on larger and less isolated world powers, such as China, India, or Russia?

In America, the loudest protests have come from commercial biotechnology companies, who argue that open-ended inspections of their labs and production facilities will leave them vulnerable to industrial espionage. Biotechnology is a multi-million-dollar industry. Between 1989 and 1996 the number of firms in the United States developing new-generation drugs soared from 45 to 113. Today's medical, industrial, and agricultural research often involves work with the same pathogens used in the development of weapons.

Some of these objections have been answered by a proposal for "managed access," which would allow the host country to negotiate the manner in which commercially sensitive labs are visited. Notified in advance of an inspection, facility managers would be allowed to partially reconfigure computers and production equipment with proprietary information. New techniques are also being developed to disrupt secret DNA sequences while allowing inspectors to detect the presence of suspicious microorganisms. Sophisticated chip-based biosensors capable of "nonintrusive" gene probes are also now on the market, but all of these have shortcomings. Nothing prevents a state from concealing a weapons program under the guise of protecting commercial secrets.

Arms treaties are important. They set standards of international behavior regarding the acquisition and use of weapons of mass destruction. But they are almost invariably ignored when countries believe their national security is at stake.

The American plan to stockpile and develop vaccines against known agents is the most comprehensive of its kind in the world. Yet as parts of that plan have been implemented, its limitations have become clear. Mandatory immunization of troops has been official Pentagon policy since 1993. All 2.3 million American soldiers have begun to receive shots against anthrax, currently regarded as the principal threat because of its documented presence in Saddam Hussein's arsenal. But no vaccines are contemplated against other agents believed to be in that arsenal, such as aflatoxin, botulinum, and smallpox. The extra cost would be enormous (the six-year anthrax vaccine program alone will cost the military an estimated $130 million) and vaccines are not without side effects. Injecting soldiers against dozens of diseases would not protect them from the agents we don't know about, or from those for which there are no known vaccines.

Vaccines work by inducing the creation of antibodies that fight specific diseases. Some are given orally, but most are injected into the muscle to insure maximum efficacy. Vaccines made of live but weakened microorganisms are generally more effective than those made of nonliving cellular or subcellular components. Both types are usually benign, but in rare cases they can trigger significant changes in the blood and endocrine systems. Some have been known to affect the functioning of the heart, lungs, kidneys, and other organs. It is not medically advisable to combine too many different courses of vaccination.

There are currently no known vaccines for brucellosis, glanders, and melioidosis or for many viral diseases, such as Ebola and Marburg. The plague vaccine was found to be ineffective against aerosol dissemination in animal studies. The tularemia vaccine is difficult to culture and potentially dangerous. Of the four possible strains available for viral encephalitis, the first and most potent (a live vaccine) produces adverse reactions in 20 percent of all cases and is ineffective in 20 percent. The second is of restricted effectiveness (it only works against three subtypes of the disease), and the third and fourth are poorly immunogenic and require multiple immunizations. The smallpox vaccine, only available in the United States to lab workers and military personnel, can be administered either before or after infection. It requires periodic boosters and wears out after ten years, though revaccination is required after three years in case of infection. Skin testing is recommended for Q fever and botulinum toxin.

The anthrax vaccine used in the United States has to be administered six times before it becomes effective (three times in two-week intervals and three times in six-month intervals) with annual boosters thereafter. Anthrax vaccines produced in other countries require different courses of inoculation. American experts maintain that annual boosters are safe—the live vaccine we used in Russia was associated with some risks—but scientists generally agree that excessive vaccination can create complications in the immune system, leading in rare cases to the formation of tumors.

Repeated vaccinations has been known to trigger or aggravate allergies. Thirty minutes after I received my last vaccination against anthrax in 1987, my face became swollen, and I developed a rash and had trouble breathing. I took Dimidrol, a powerful anti-allergy medication available in Russia (though not in the United States) and felt better again in a few hours. For the next ten days I received intravenous preparations at a hospital—a form of allergy treatment we called desensibilization therapy. Several colleagues had been forbidden from anthrax work after similar reactions. I knew this was a sign that I was genetically susceptible to large quantities of specific foreign proteins, and that my immune system had been stretched to its limit. I received my first anthrax vaccination in 1979 and began a course of annual vaccinations in 1982. I was also vaccinated against smallpox once, twice against tularemia, and four times against plague. The chronic allergies I have suffered throughout my adult life are a direct consequence of repeated exposure to live vaccines, and to other biological substances I worked with.

Vaccines provide excellent protection against specific diseases, but the characteristic that makes them so effective—that specificity—is also the source of their limitations. Smallpox antibodies offer no protection against plague. A typhoid vaccine will not lower the risk of measles. Combined vaccines are possible, such as the diphtheria-pertussis-tetanus shot given to children, but most of these go straight to the metabolism of specific organisms. A vaccine works against a single pathogen, or occasionally several similar ones, but an all-purpose antidote does not exist.

The use of vaccines for biodefense makes sense when we know what agent is likely to be used and when we can identify a specific

target population—troops, for instance, within range of a known
arsenal. But the protection they confer must be measured against a
shifting threat. An adversary who knows that his opponents'
troops have been inoculated against anthrax can switch his battle
plans to smallpox or plague—or to an agent for which no vaccine
exists. We can vaccinate our soldiers against a minimal combination of the most likely threats, but we will still not know whether
an opponent has developed a weapon virulent enough to overcome
existing antidotes.

Despite American efforts and expenditures, vaccines have limited value for the protection of civilians. Who would be deemed
vulnerable? And which agents should they be protected against? A
crash program to increase the available doses of smallpox vaccine
in the United States (currently seven million) might deter a country
or a terrorist group from launching a smallpox attack, but there
are plenty of other options. And who would get those seven million doses if several cities are attacked at once? The city of New
York alone has a population of over seven million. Will each city
have its own stockpile?

I am not suggesting we should drop vaccines from our biodefense plan, only that we should keep their effectiveness in perspective. Even if we could afford the expensive and lengthy process of
development, testing, and approval currently required for the in-
troduction of new vaccines in the United States and most Western
countries, the continued advances in weapons-making knowledge
will always put us a step behind.

Over the past two decades, scientists have vastly expanded our understanding of how the immune system works. This knowledge
can be exploited to provide a new form of medical defense against
biological agents. In the simplest terms, the immune system works
by distinguishing our cells from the alien microorganisms that invade our bodies every day. We have at our disposal a network of
agents programmed to make such distinctions and report on their
findings. New antibodies are continuously being formed to recognize specific threats and eliminate them before they cause damage.
These antibodies and the agents that code for their formation are
endowed with what we call memory—the ability to recognize a

previous invader. This subcellular capacity lies at the heart of the success of vaccines. For years immunologists focused exclusively on vaccines and the antibodies they produced—the most visible elements of specific immunity—ignoring the processes that have come to be grouped together under the general rubric of nonspecific immunity.

One of the first researchers to observe nonspecific immunity in action was a Russian microbiologist named Ilya Mechnikov, who identified the first cellular components of immunology. Working in Italy between 1882 and 1886, he noticed that a collection of cells would migrate to the site of infection, where they would surround, swallow, and destroy foreign particles. Mechnikov called these cells phagocytes (they are now also referred to as macrophages or monocytes). His work, which earned him the Nobel Prize for Medicine in 1908, laid the foundation for the modern science of immunology.

It was not until the 1960s that scientists began to focus on the cells and molecules responsible for coordinating the body's nonspecific immune response against invaders. These include macrophages and granulocytes, as well as special proteins in the blood that interact to defeat foreign microorganisms in what is called a complement cascade. Another important component of nonspecific immunity is a remarkable group of molecules called cytokines, through which cells communicate to one another and pass on vital marching orders.

Cytokines form a bridge between specific and nonspecific immune systems. They are produced in response to viruses or bacteria, or to a general stimulus in the blood. Their function is primarily regulatory: they direct the magnitude of an immune response. They can suppress or stimulate the secretion of antibodies and macrophages, induce fever and inflammation, and prompt the growth and activation of a host of essential immune cells. Most cannot kill viruses or bacteria on their own, but they have been found to boost the immune system and to enable patients to respond to previously unproductive treatments. They have also been shown to increase the effectiveness of T and B lymphocytes including natural killer cells, which destroy pathogenic bacteria and cells invaded by viruses.

In 1957, European scientists identified the first cytokine. It was named interferon, because it appeared to interfere with the progress of viral infections. Three major types have been identified. Interferon took years to isolate, but by 1979 scientists at an American pharmaceutical firm had managed to reproduce one type—interferon alpha—artificially. Touted as the "antiviral penicillin," interferon entered the medical lexicon as a powerful tool for the treatment of illnesses ranging from hepatitis to Kaposi's sarcoma, a frequent symptom of AIDS. Scientist have since become more cautious in their claim, as interferon produced mixed results in lab tests and was found to cause side effects when taken in large doses. Nevertheless, it is widely used today.

The discovery of cytokines and other elements of nonspecific immunity represent an important step forward for medicine. Scientists in America have developed a treatment for AIDS that includes interleukin-2, another cytokine, and research into the effects of cytokines on tuberculosis and other diseases is under way in the Netherlands, Great Britain, Japan, France, and Canada. At least eighteen interleukins are well known to scientists today, and each year more are discovered.

Nothing will replace the long-term protection provided by vaccines against specific diseases, but boosting our nonspecific immune system may offer at least temporary protection from pathogenic agents and possibly could go even further. If administered in the crucial first hours after an attack—when authorities are still trying to identify which agent was used and organize a medical response—such a booster could help contain the crisis. It is a long shot, but everything I know about biological weapons tells me that this is far more promising than attempts to rig office buildings and public monuments with detection devices or to stockpile vaccines.

Ten years, almost to the day, after I was called to army headquarters in Moscow to brief Soviet colonels on how to load intercontinental missiles with anthrax and plague, I met with two U.S. Marine colonels in the fifth-floor conference room of an office building in Virginia.

The marines had driven up from their training base at Quantico, where they operate a think tank called the War-Fighting Lab. Despite the name of their lab, the marines came to discuss defense:

how to protect troops against biological warfare and terrorism. Often the first on the scene in a military emergency, marines are exposed to the kinds of unconventional threats not faced by other branches of the armed forces.

On May 20, 1998, I had presented to the U.S. Congress a proposal for developing nonspecific immunological defense against biological weapons. At the time, national efforts were devoted almost exclusively to detection and vaccination—a week later President Clinton would propose a reserve stockpile of vaccines—and this unconventional approach was greeted with widespread skepticism. But things changed dramatically over the next six months.

In December 1998, a committee of scientists appointed by the National Research Council's Institute of Medicine and chaired by Peter Rosen, director of the emergency medicine residency program at the University of California's school of medicine in San Diego, proposed that new research be undertaken into "broad-spectrum anti-bacterial and anti-viral compounds" to counter biological and chemical terrorism—in other words, nonspecific protection against a variety of biological weapons. This recommendation was only one of sixty projects identified by the committee, but it was singled out as a high priority. Endorsed by a panel of twelve prominent U.S. scientists, including Dr. Donald Henderson, one of the architects of the worldwide campaign against smallpox, and the Nobel prize–winning biologist Joshua Lederberg, it represented the first time such an idea had received professional review anywhere in the world.

The marines learned of my proposal before the panel delivered its findings. A congressional aide told them about my testimony before the Joint Economic Committee hearing on terrorism, and a meeting was arranged at the offices of the scientific research and development company where I now work. The two colonels took notes as I went over my ideas. I couldn't help but notice the irony of the situation. To these men I was just another civilian, a scientist with an interesting proposal.

A month later, in November, the marines called my office to report that they had received preliminary approval from their superiors to test a program of nonspecific immunity. Plans for a pilot project are under way.

In helping my adopted country create a new system of defense against the weapons I once made, I often remember Russia, which I loved and continue to love. I want this country to have a different fate. In an interview with a Russian paper, one of my friends called me a betrayer. I am certain he is not alone in thinking this. As I return to this question again and again, I have come to the conclusion that I did not betray Russia so much as it has betrayed its people. As long as it makes heroes of the people who create prohibited weapons, as long as it continues to help foreign dictators who murder civilians and to wage wars against its own people, as long as it trains its physicians and teachers to kill and considers as criminals those who try to speak against this—to call what is immoral by its name—as long as this continues, there can be no hope for a better future. We talk about economic and structural reform, but what is needed in Russia is moral reform, and until that happens, Russia will not change.

As a young boy in Kazakhstan I once came across a book about a doctor who risked his life and health to heal his patients. He was the physician I dreamed of becoming. I cannot unmake the weapons I manufactured or undo the research I authorized as scientific chief of the Soviet Union's biological weapons program; but every day I do what I can to mitigate their effects. The realization that even today, in Iraq or China, another father of three may be sitting down at a conference table to plot the murder of millions of people is what spurs me on. This is my way of honoring the medical oath I betrayed for so many years.

THE SOVIET BIOLOGICAL WARFARE SYSTEM (1990)

Appendix 1

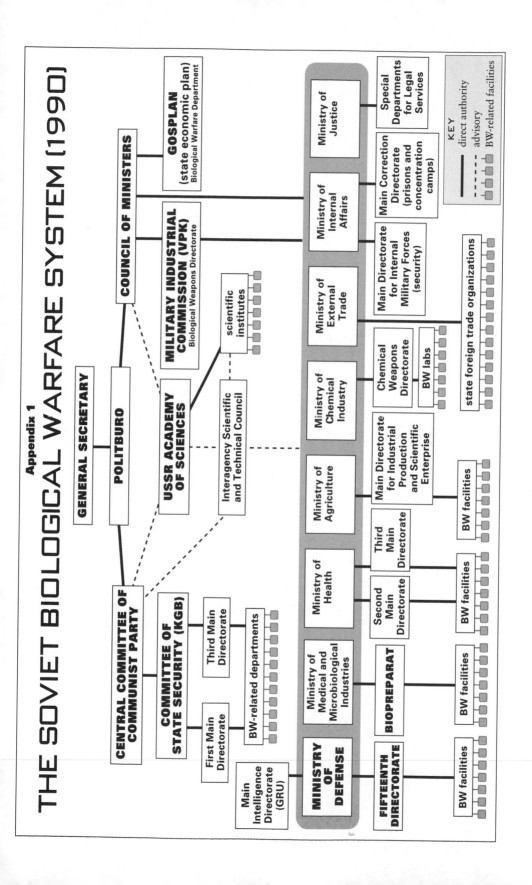

SOVIET BIOLOGICAL
WARFARE SYSTEM

Soviet Agencies, Institutions, and Facilities Involved in
Biological Weapons Research, Development,
and Production, 1973–1990

This is the first comprehensive overview of the Soviet biological weapons system. The description of the services and facilities run by Biopreparat; the Ministry of Heath; the Academy of Sciences; and the Ministries of Internal Affairs, Justice, and External Trade are complete. I know only indirectly of two to three additional facilities run by the Ministry of Defense and the Ministry of Agriculture. I was never told of those run by the KGB. In instances in which I was uncertain as to the exact name of an institute, I have given only its location. None of this information has ever been made public before.

CENTRAL COMMITTEE OF THE COMMUNIST PARTY

The central authority of the Soviet Communist system, comprising Party representatives elected from around the country. Party leaders sat on the Politburo, the Party's governing body. The general secretary of the Party was chairman of the Politburo and effectively leader of the country. One member of the Politburo was assigned responsibility for all military and defense industry issues, including biological warfare.

The Central Committee Secretariat acted as the Party's administrative apparatus. The secretariat's Defense Department was in charge of biological warfare as well as other military and defense-related matters. The Committee of State Security (KGB) was the Party's intelligence and security organ. Its chairman sat on the Politburo.

COUNCIL OF MINISTERS

The highest body of the Soviet government, consisting of the prime minister (the de jure head of government) and all government ministers. Ministers derived their real authority in the system from their positions in the Party. The most powerful ministers, including the prime minister, had seats on the Politburo and were thus subordinate to the Party organs and to the general secretary.

Gosplan, the state economic planning agency, was attached to the Council of Ministers and authorized budget expenditures for all government departments and activities, including Defense (but it had no authority over Party spending). Gosplan's Biological and Chemical Weapons Department supplied funding to all biological weapons (BW) facilities. The State Technical Commission, based in Moscow, developed and supervised electronic surveillance and electronic counterespionage measures at all biological weapons facilities.

Ministers responsible for defense-related industries sat on the Military-Industrial Commission, which ran the Soviet Union's military-industrial complex. The chairman was a deputy prime minister, but the commission's activities were ultimately responsible to the Politburo member in charge of military matters. The Military-Industrial Commission was divided into directorates responsible for different military sectors. The Biological Weapons Directorate coordinated the development and production of biological weapons.

MINISTRY OF DEFENSE

The General Staff Operations Directorate of the Ministry of Defense was the chief war-fighting organ. The Special Biological Group, attached to this directorate, was responsible for the development of biological war-fighting doctrine and logistics. The Main Intelligence Directorate (GRU) was responsible for military espionage and counterespionage. GRU agents conducted covert operations abroad to monitor foreign biological weapons programs and acquire strains of pathogens and documentation that might be useful for the USSR's biological weapons program. Biological weapons procurement (including testing, approval of weapons for the Soviet arsenal, and annual quotas for biological weapons stockpiles) was supervised by the Special Armaments Group, a unit administered by the deputy minister of defense for armaments.

The Fifteenth Directorate of the Soviet army developed and produced biological weapons. It commanded specific military units—some of brigade size—assigned to testing grounds and to provide security for military biological weapons facilities. In the event of war, biological weapons deployment would be the responsibility of military units of the Strategic Rocket Forces and Air Force. Between 1945 and 1973, the Fifteenth Directorate was the leading Soviet agency for biological weapons research. Although its research and planning role was eclipsed by the creation of Biopreparat, it continued to control biological weapons stockpiles and the primary facilities for biological weapons production. The following facilities were under its control.

Zima, Irkutsk region, Railway Station Zima. Storage, anthrax weapons.
Kirov, Institute of Microbiology. Developed BW weapons: typhus, Q fever, tularemia, brucellosis, glanders, anthrax, and melioidosis. Studied toxin weapons. Researched genetically altered bacterial strains. Produced and stockpiled plague.
Moscow region, Kubinka Military Airport. Home base for Air Force unit that delivered cargo, personnel, and animals to testing grounds.
Moscow, Institute of Safety Techniques. Developed BW production equipment.

Nukus, Kara Kalpak Autonomous Republic. Testing ground for BW and chemical warfare simulants.

Reutov, Moscow region. Storage: BW warheads, bombs, and bomblets.

Shikhany, Volga River region. Testing ground for chemical weapons and BW simulants.

Strizhi, Kirov region. Manufactured viral and bacterial BW. Built in the late 1980s, it was the last plant created by the Fifteenth Directorate before the collapse of the USSR.

Sverdlovsk (now Yekaterinburg), Institute of Military Technical Problems. Developed BW: anthrax, tularemia, glanders, melioidosis. Researched toxic weapons, including botulinum toxin. Researched antibiotic-resistant anthrax and multidrug-resistant glanders. Produced anthrax, glanders. Stockpiled anthrax.

Volga River region, Air Force base (exact location unknown). Bomber base that may have been used as launchpad for aircraft disseminating BW during Afghan war.

Vozrazhdenie (Rebirth) Island, Kazakhstan. Testing grounds for BW. Command and control center located in nearby city of Aralsk.

Zagorsk (now Sergiyev Posad) Virology Institute. Researched and developed smallpox, monkey pox, Bolivian hemorrhagic fever, Argentinean hemorrhagic fever, Marburg, Ebola, Lassa fever, Rift Valley fever, Venezuelan equine encephalitis. Researched Japanese encephalitis, tick-born encephalitis, Eastern equine encephalitis, Western equine encephalitis, Murray Valley encephalitis, Saint-Louis encephalitis, and others. Produced and stockpiled smallpox.

MAIN DIRECTORATE "BIOPREPARAT"

Created in 1973 to provide civilian cover for advanced military research into biological weapons, the agency was originally attached to the Council of Ministers. The majority of its personnel came from the army's Fifteenth Directorate, which kept it effectively under military control. A government reorganization in the mid-1980s transferred Biopreparat to the Ministry of Medical and Microbiological Industries, but it continued to enjoy virtually autonomous authority as the principal government agency for biological weapons research and development. Biopreparat was officially responsible for civilian facilities around the country dedicated to the research and development of vaccines, biopesticides, and some laboratory and hospital equipment; but many of its facilities doubled as BW development and production plants and were earmarked as "reserve" or mobilization units for use in case of war.

Berdsk, Technological Institute of Biologically Active Substances. Developed enzymes for research into genetically altered biological weapons.

Berdsk, Scientific and Production Base (Siberian Branch of the Institute of Applied Biochemistry). In operation from 1975 to 1981, it developed filling and assembling techniques for BW production, as well as lab techniques for production of brucellosis BW.

Berdsk, production plant. Mobilization (reserve) facility: plague, tularemia, glanders, and brucellosis. Target capacity up to 100 tons of each weapon annually.

Kirishi, Leningrad region, Special Design Bureau for Precision Machinery Building. Developed equipment for manufacturing and testing BW.

Koltsovo, Novosbirisk region, Institute of Molecular Biology "Vector." Researched and developed viral weapons: smallpox, Ebola, Marburg, Bolivian hemorrhagic fever, Venezuelan equine encephalitis, Russian spring-summer encephalitis. Studied numerous other viruses for possible BW use, including HIV. Developed genetically altered viruses for potential weapons use. Developed new production techniques for making smallpox and Marburg weapons. Researched genome of viruses in order to create "chimera" (combined) viral weapons.

Kurgan, Combine "Syntez." Mobilization (reserve) plant for manufacturing a liquid form of anthrax biological weapon. Assigned target capacity: 1,000 tons of unmodified anthrax weapons over one year.

Leningrad (now St. Petersburg), Institute of Ultra-Pure Biopreparations. Researched and developed techniques for testing and application of BW. Studied application of BW to cruise missiles.

Lyubychany, Institute of Immunology. Researched biological agents used to suppress human immune system.

Moscow, Institute for Biological Instrument Design. Developed BW detection equipment, biosafety equipment, and biosafety procedures.

Moscow, Institute of Applied Biochemistry. Designed BW manufacturing and testing equipment. Developed standards for industrial-scale manufacturing of biological weapons.

Moscow, Design Institute "Giprobioprom." Designed BW research and production facilities.

Obolensk, Moscow region, Institute of Applied Microbiology. Researched and developed plague, tularemia, glanders, anthrax. Developed drug-resistant and vaccine-resistant weapons (Metol), and weapons targeted on central and peripheral neural systems (Bonfire).

Omutninsk, Scientific and Production base (formerly Eastern European Branch of the Institute of Applied Biochemistry). Researched and developed plague, tularemia.

Omutninsk, chemical production plant. Mobilization (reserve) facility for manufacturing plague, tularemia, glanders. Target capacity up to 100 tons of each weapon annually.

Penza, Combine "Biosyntez." Mobilization (reserve) plant for manufacturing dry form of anthrax BW. Assigned target capacity: 500 tons of unmodified anthrax weapon over one year.

Stepnogorsk, Kazakhstan, Progress Scientific and Production Base (formerly Kazakhstan branch of the Institute of Applied Biochemistry). Mobilization (reserve) facility designated to produce 300 tons of modified form of anthrax BW over 250 days. Also designated for production of plague, glanders, and tularemia BW. Researched and developed anthrax and glanders. Tested anthrax, glanders, Marburg.

Vilnius, Lithuania, Institute of Immunology. Researched and developed enzymes for molecular biology and genetic engineering research. The research was subsequently used to develop genetically altered weapons at other facilities without the knowledge of institute officials.

Yoshkar-Ola, Mordovia Autonomous Republic, Special Design Bureau of Controlling Instrument and Automation. Designed and manufactured equipment and instrumentation for BW weapons development, testing.

MINISTRY OF AGRICULTURE

Almaty, Kazakhstan, Production Facility "Biocombinat." Reserve mobilization plant for production of BW, primarily anthrax.

Golitsino, Scientific Institute of Phytopathology. Developed anticrop weapons, including agents for destruction of wheat, rye, corn, and rice.

Otar Railway Station, Kazakhstan, scientific institute and test site. Tested anticrop and antilivestock BW.

Pokrov, production plant. Mobilization facility for manufacture of smallpox (up to 100 tons) and Venezulelan equine encephalitis (40 to 80 tons), as well as other viral weapons. Reserve facility for antilivestock BW.

Tashkent, Uzbekistan, Scientific Institute of Phytopathology. Researched and developed anticrop weapons.

Vladimir, research and development facility. Researched and developed antilivestock weapons: African swine fever, foot-and-mouth disease, rinderpest, etc.

MINISTRY OF CHEMICAL INDUSTRY

Several labs operating under the control of the Chemical Weapons Directorate were involved in biological weapons work, including the development of toxic organic substances. At least one lab was located in Moscow.

MINISTRY OF HEALTH

Second Main Directorate

Controlled about a dozen antiplague institutes and research facilities for work on microbiology and epidemiology scattered around the USSR. In addition to peaceful medical research, overseen by the Main Sanitary Epidemiological Directorate, which ran them, these facilities were responsible for investigating new strains of pathogenic microorganisms that could be used as biological weapons.

The institutes were located in Minsk (Belarus), Saratov, Irkutsk, Samara (formerly Kuybyshev), Rostov-on-Don, Almaty, and Volgograd, among others.

Third Main Directorate

Controlled a network of special hospitals and medical units to serve biological weapons research-and-development facilities. A second network investigated biological agents that could cause nonlethal and lethal organic and physiological changes (Program "Flute"). Several labs in this second network developed toxins and other substances for use against "individual human targets."

Moscow, Medstatistika. Gathered medical/BW intelligence from all over the world, mostly from open-source medical journals, but also analyzed information gathered covertly.

Moscow, Institute of Applied Molecular Biology (later Russian Scientific Center of Molecular Diagnostics and Treatment). Studied various biological substances in order to find those that could kill or cause irreversible mental damage.

Moscow, Institute of Immunology. Studied regulatory peptides with toxic properties capable of triggering both reversible and irreversible changes in the neural and immune systems.

Moscow, Scientific and Production Center of Medical Biotechnology. Basic research into human genome, in order to identify new BW possibilities.

Moscow Region, Center of Toxicology and Hygienic Regulation of Biopreparations. Studied toxic biological compounds with a high killing potential for aerosol use.

Sukhumi, Georgia. Breeding facility for monkeys used in BW testing.

USSR ACADEMY OF SCIENCES

The state scientific federation of the Soviet Union controlled funding and organized research in every major scientific discipline. Leading scientists accepted for membership earned the title of Academician (Akademik). The academy acted in an advisory capacity to the Council of Ministers and the Central Committee. The Inter-agency Scientific and Technical Council, formed in the early 1970s, coordinated advanced research in biological weaponry. The chairman was a government minister. Members included ranking representatives from the Central Committee, the Fifteenth Directorate, and Biopreparat; directors of leading scientific institutes; the vice president of the Academy of Sciences; the first deputy minister of health; the deputy minister of chemical industries; and the chief of the biological warfare directorate of the Ministry of Agriculture. The council acted as the principal scientific and industrial advisory body to the biological weapons system.

Moscow, Institute of Bioorganic Chemistry. Fundamental research into BW.

Moscow, Institute of Molecular Biology. Fundamental research into BW.

Moscow, Institute of Protein. Fundamental research..

Moscow, Institute of Biochemistry and Physiology of Microorganisms. Fundamental research.

Vladivostock, Pacific Ocean Institute of Bioorganic Chemistry. Fundamental research into marine toxic substances..

COMMITTEE OF STATE SECURITY (KGB)

First Main Directorate

Responsible for foreign intelligence gathering, including the monitoring of foreign biological weapons programs. Conducted its own research into biological weapons, primarily for assassination purposes. Controlled several covert research units for chemical warfare and biological weapons, including Laboratory 12.

Third Main Directorate

Responsible for domestic counterintelligence and security. Regional branches provided security for individual Biopreparat facilities as well as camouflage and disinformation operations.

MINISTRY OF INTERNAL AFFAIRS

Moscow, Main Directorate of Labor-Correction Enterprises. Supervised prisons and concentration camps. Provided prison labor for the construction of BW facilities.

Moscow, Main Directorate of Internal Military Forces. Provided guards for BW facilities not controlled by the army's Fifteenth Directorate.

MINISTRY OF EXTERNAL TRADE

Responsible for Soviet foreign trade. Special departments arranged covert purchases of equipment and animals used in biological weapons program. Representatives and agents were posted abroad.

MINISTRY OF JUSTICE

A special department at the Ministry of Justice was responsible for legal services for biological weapons facilities and personnel. It included special prosecutors, lawyers, judges, and special courts.

ACKNOWLEDGMENTS

I want to thank Stephen Handelman, who provided me with the voice to tell my story, using his excellent writing skills. I benefited from the added dimension he brought to this book with his knowledge of Russia, but perhaps most of all from his friendship.

Special thanks are due to Joy de Menil, my editor at Random House, whose probing mind and dedication kept us on our toes from start to finish. Much of this book is due to her vision. I am also grateful to Jennifer Guernsey, whose research skills, knowledge of biology, and enthusiasm were invaluable. Both Stephen and I are in her debt. Cynthia Cannell, my agent, deserves special gratitude for perceiving the importance of this project early on and for her tireless marketing work.

Particular thanks are due to Charles Bailey and Bill Patrick, who have provided—and continue to provide—support and advice to a scientific colleague once separated from them by Cold War politics. I am grateful to Admiral Elmo Zumwalt, Jr., for his encouragement and to Vaughan Forrest, a senior staffer with the Joint

Economic Committee of Congress, who was among the first to perceive the potential of nonspecific immune defense.

I want to thank Frank Cilluffo, Walter Dorn, Ambassador Jack Matlock, Jr., Susan Simpson, and Jessica Stern for their insight and assistance at various stages of this project. I am grateful to Hal Pastrick and other colleagues at SRS Technologies for initiating me into the ways of American research and defense and to my new colleagues at Batelle Memorial Institute for providing support with my new scientific projects.

The friendship and support of Olga Deviatkova, Mark Berkovich, Melissa Bailey, Jenny Patrick, Harry H. Horning, and Marat and Sasha Akchurin have made my integration into life in America less difficult than it might otherwise have been. This book is a testament to their support. I must also thank those friends and officials in Russia, Kazakhstan, and the United States who assisted my family in its passage to America. Although I cannot name them here, these pages are a way of expressing my heartfelt appreciation for the risks they took. I wish especially to thank my friend Victor Seemann, a former intelligence colonel (the only one whose name I can mention), whose help in understanding American life was invaluable.

To Lena, the mother of my lovely children, I owe a debt that can never be repaid. Her love and counsel carried over from our life together in the Soviet Union to our resettlement in a new and confusing world. Mira, Alan, and Timur have been a source of constant inspiration to me. As I watch them blossom and grow in the peaceful society they've made their own, I realize they are the real reason for writing this book.

Finally, I wish to dedicate these pages to my parents, as well as to my brother, Bakhit, and my sister, Saule. May their love continue to strengthen me for many years to come.

INDEX

ABOUT THE AUTHORS

KEN ALIBEK was born in Kauchuk, Kazakhstan, in 1950. He graduated from the military faculty of the Tomsk Medical Institute in 1975, where he majored in infectious diseases and epidemiology. He holds Ph.D.s in microbiology for research and development of plague and tularemia biological weapons and in biotechnology for developing the technology to manufacture anthrax biological weapons on an industrial scale. He joined Biopreparat in 1975 and was deputy chief of the agency from 1988 to 1992. Since he defected to the United States in 1992, he has briefed U.S. military intelligence, and he is now working on biodefense.

STEPHEN HANDELMAN is a columnist at *Time*. He was Moscow bureau chief for *The Toronto Star* in the late 1980s and early 1990s. He is the author of *Comrade Criminal: Russia's New Mafiya*.

ABOUT THE TYPE

This book was set in Sabon, a typeface designed by the well-known German typographer Jan Tschichold (1902–74). Sabon's design is based on the original letterforms of Claude Garamond and was created specifically to be used for three sources: foundry type for hand composition, Linotype, and Monotype. Tschichold named his typeface for the famous Frankfurt type-founder Jacques Sabon, who died in 1580.